Cordon Bleu

Home Freezing

Cordon Bleu

Home Freezing

CBC / B.P.C. Publishing Ltd.

Published by
B.P.C. Publishing Ltd.,
St. Giles House, 49/50 Poland Street,
W1A 2LG

Copyright B.P.C. Publishing Ltd., 1972

Designed by Melvyn Kyte
Printed by Waterlow & Sons Ltd., London and Dunstable

These recipes have been adapted from the Cordon Bleu Cookery Course published by Purnell in association with the London Cordon Bleu Cookery School Principal: Rosemary Hume. Co-principal: Muriel Downes

Quantities given are for 4 servings. Spoon measures are level unless otherwise stated

Contents

Introduction	page	7
Choosing a Home Freezer	page	9
How Deep Freezing Works	page	13
Packaging—How and Why	page	14
How Long will it Keep?	page	19
Thawing It Out	page	23
The Basic Rules	page	25
Freezing Uncooked Food	page	26
Freezing Recipes	page	31
Appendix	page	137

Introduction

A home freezer brings a new dimension to modern cookery. It is not just a device for farmers' wives who have surplus produce to store, or for the wife who goes out to work and so does not have the time to shop frequently in the normal way; it is a wonderful aid to anyone who cooks. Not only does it enable you to freeze fresh food when it is plentiful and cheap, but you can cook for the future and confidently lay down soups and sauces, pâtés and puddings, pies and pastry or stocks and breadcrumbs, even whole meals ready to be heated and enjoyed months from now. And you still have the satisfaction of knowing that what you will be eating will have been all your own work.

With the help of the home freezer mothers can do more of their cooking during the week when the children are at school and have to spend less time in the kitchen during weekends and holidays. Party dishes can be deep-frozen, leaving the hostess time to relax and entertain her guests instead of worrying about progress in the kitchen.

Commercially-frozen food, while not always as cheap as some suppliers would have us believe, is useful. Home-frozen fresh food bought at glut prices can be a great money-saver. Food that you have prepared at home and then put in the home freezer will make your friends and guests realise that there is no 'lazy cook' label to be pinned on you.

Of all the methods of preserving food that have been invented so far the deep-freeze is the most successful and is rapidly becoming the most extensively used. Providing you stick to the rules there is no reason why any project should fail. Freezing cannot improve the quality of food but it does provide the most effective means yet devised of halting the chemical action that turns food first stale and then bad. It suspends the growth of micro-organisms that cause this chemical action. Admittedly there are a few types of food that do not take kindly to being frozen but these are very few compared with the wide range that lend themselves to the process which quickly grips them in a state of suspended goodness and delicious flavour and appearance and keeps them that way in a form of storage whose cost is not only low but is also amply compensated by the convenience of being able to cook in bulk and plan well ahead.

Rosemary Hume
Muriel Downes

Choosing a home freezer

Before you embark on the business of deep-freezing, give a good deal of thought to the kind of freezer you want. This will depend on a number of factors, including the size of your household, the type of life you lead, whether you intend to use it mainly for storing food to be cooked or have plans for home-freezing your own cooking for use at a later date.

There are two main types of freezer, the upright style with the door or doors on the front and the chest type, with a lid that opens upwards. It is a scientific fact that cold air tends to fall downwards and warm air rises, so that in the case of the upright model some cold air will fall out each time the door is opened. When the cold air falls out it is automatically replaced by warm air at the top and that warm air will almost certainly contain some moisture, which in time will condense inside the freezer. This means that this type of freezer will need defrosting three or four times a year, but it does have the advantage that the food is more accessible. Chest type freezers release less of the cold air when opened but it is less easy to see what is in the freezer, and to reach the bottom you have to reach in and lean over—not the best posture for lifting or work of any sort. This type of freezer usually needs defrosting only once a year.

The size of the freezer will depend entirely on how you want to use it and what your storage requirements will be. Sizes range from those giving about 1 cubic foot of storage space to the large commercial types of 16 cubic feet or more. Many of the very small types are combined in one cabinet with an ordinary refrigerator. But do not confuse these with the freezing part of an ordinary fridge. It is not the same thing. Few people would find less than 4 cubic feet adequate for their needs, although of course to hold just enough commercially frozen food in a town flat or small country weekend cottage the smaller ones would be sufficient. Nor is it easy to visualise a household of only two people needing 16 cubic feet of storage space. Some people who have a lot of spare produce find it more convenient to hire a locker in a deep-freeze plant and keep only a small freezer at home. But it is also a fact that the big freezers are relatively cheap to buy because the purchase price does not increase in proportion to the capacity. Again, against this possible economy, unless you are going to keep the freezer full you will find yourself paying money every week just to keep some air cold, the air being that part of the freezer capacity which is not filled with food.

As a rough guide, you can reckon on being able to put between 20 and 25 lb of food into each cubic foot. If the food is in space-saving commercial packs you might get as much as 30 lb in. If you allow about 2 cubic feet per person and an odd couple of feet over you should be near enough to the mark for your needs.

One important factor that goes with the size of the freezer

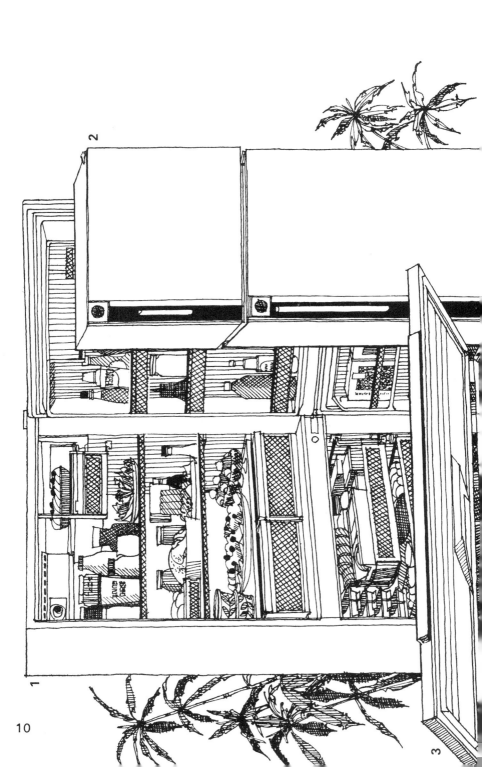

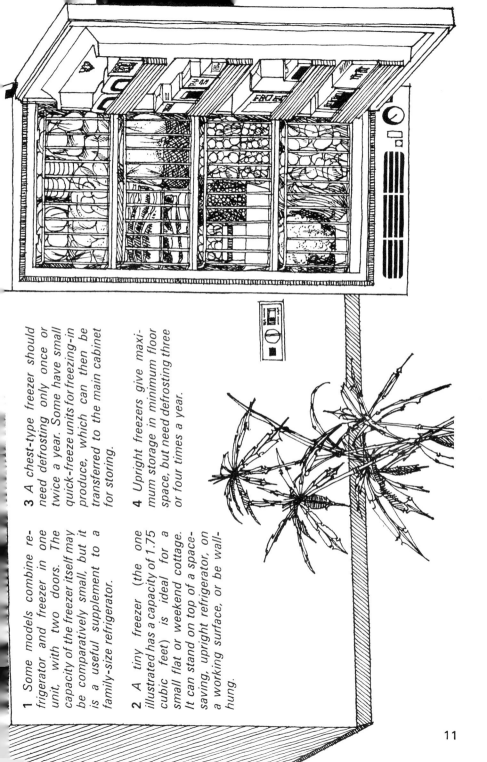

1 Some models combine refrigerator and freezer in one unit, with two doors. The capacity of the freezer itself may be comparatively small, but it is a useful supplement to a family-size refrigerator.

2 A tiny freezer (the one illustrated has a capacity of 1.75 cubic feet) is ideal for a small flat or weekend cottage. It can stand on top of a space-saving, upright refrigerator, on a working surface, or be wall-hung.

3 A chest-type freezer should need defrosting only once or twice a year. Some have small quick-freeze units for freezing-in produce, which can then be transferred to the main cabinet for storing.

4 Upright freezers give maximum storage in minimum floor space, but need defrosting three or four times a year.

Choosing a home freezer continued

is its capacity for freezing quickly the dishes that you have prepared yourself. The faster you freeze the better, and the lower the temperature you can attain the faster you will freeze. Most of the existing models will freeze up to one tenth of the freezer's total capacity in 24 hours, but as more modern models come off the drawing board they have extra gadgets for super-quick freezing. If you try to freeze too much food in a day you will find that instead of fast freezing you are getting only a slow freeze and, as shown elsewhere in this book, that does not always do the job. Another important point in choosing your freezer is the availability of service facilities. If anything should go wrong your freezer would need servicing in a hurry and you want to be sure that there is an engineer in your area who can do the job.

With increasing competition in the market there isn't really much to choose in terms of price, but you can shop around and sometimes save perhaps 15 or 20 per cent of the price by buying in a sale, or a marked-down last year's model. Beware of some models which are in effect only glorified ice cream conservators; while they will store food for a certain time they will not have the fast-freeze equipment or sufficiently low temperatures to do what you want.

If you should want to figure out what you will be paying, reckon the life of your freezer at 10 years and divide this into the total cost of the freezer, including any hire purchase charges if you are buying it that way. You should reckon on about £3 a year for servicing and the electricity you will use should not be more than about $1\frac{1}{2}$ or 2 units per cubic foot capacity per week, whatever the charge per unit is in your area. Allow about 1p per lb of frozen food for packaging material. You can also insure the contents of your freezer against loss due to accidental power failure which could cause the frozen food to thaw out and be spoiled. Elsewhere there are instructions about how to meet such a threat if it should ever happen, but for costing purposes you can allow about 5p for every £1 worth of food in the freezer to cover the insurance premiums.

Apart from the basic cost of the freezer there are some optional refinements that may push up the cost but which might be worth your while considering. They include locks, a device to show when the temperature inside the cabinet has risen, separately refrigerated pull-out drawers, door shelves on upright models, work tops on the chest freezer or a super-fast freeze switch for extra-fast freezing.

How deep freezing works

Food begins to deteriorate from the moment it is slaughtered, in the case of meat, poultry or fish, or from the time it is picked, in the case of fruit or vegetables. This is because of the chemical reaction of micro-organisms such as enzymes, moulds, yeasts and bacteria. In the living plant or animal some of these helped the living cells to grow and multiply, but nature provided for the destruction of such matter by turning about the purpose of these micro-organisms after death. The destructive power of these micro-organisms can be halted by extreme heat in processes such as canning, but these and other methods of food preservation leave a great deal to be desired because they cause a permanent alteration in the taste, flavour and appearance of the foods concerned, usually for the worse. At the other end of the temperature scale, freezing at very low temperatures keeps the micro-organisms in check without undue effect on the texture, colour and flavour of the food, virtually holding it in a state of suspended animation.

At this point it should be emphasised that the domestic refrigerator does not have the freezing capability needed for deep-freezing. The cabinet of this type of refrigerator generally is kept at something like 40°F (4°C), whereas water freezes at 32°F (0°C). Some fridges with three-star frozen food compartments get the temperature down to 0°F ($-18°$C) but although this might be sufficient for storing some frozen foods for relatively short periods, it does not get down to the $-12°$F ($-24°$C) at which fresh food should be deep-frozen or the $-30°$F ($-34°$C) at which foods are deep-frozen commercially.

Broadly speaking the quicker a food is frozen and the slower it is thawed out the less change there will be in texture, flavour and appearance. The lower the temperature to which it is reduced and kept, the longer it will remain good since the greater will be the effect on the activity of the micro-organisms. Also the faster the freezing process the finer will be the ice crystals formed inside the food. This means a lessening of the possible damage to cells in the food on thawing. Most home freezers now sold have a special section for freezing the food you have prepared and a super-fast switch to lower the temperature to that required. The average freezer can freeze about 10 per cent of its total capacity in 24 hours in this freezing section and this food is then moved over to the normal storage section to make room for more food to be frozen if needed. Freezing does not eliminate the effect that oxygen in the air has on food and therefore the food must be sealed off against the air. This is done by packaging in plastic film, foil and airtight cartons as shown in the special section on packaging.

In some foods enzyme activity continues even when the food is deep-frozen. In the case of most vegetables this activity has to be further checked by blanching, immersing in boiling water for very short periods before

How deep freezing works
continued

deep freezing.

Among the few foods that cannot be successfully deep-frozen are:

Egg-based sauces such as mayonnaise. These tend to separate when thawed out, but are successful if incorporated in other dishes.

Salad vegetables such as cress, lettuce, chicory, etc. They go mushy and frostbitten. Celery is not a success.

Fresh tomatoes do not freeze well; reduce them to a purée first.

Potatoes, particularly new potatoes, are not successful.

Root vegetables should not be frozen unless very young.

Melons, avocado pears, bananas and some other fruits not well known in Britain should not be deep-frozen.

Cream may be deep-frozen only if it contains more than 40% butter-fat.

Hard-boiled eggs become very rubbery; fresh eggs can only be deep-frozen successfully if separated and 1 teaspoon salt added to each 4 egg yolks.

Shellfish usually require a deeper freeze than is obtainable at home.

Packaging - how and why

Every piece of food that goes into a freezer must be properly wrapped before it is frozen. This is not a manufacturers' gimmick to sell more packaging materials. There are very good scientific reasons for it.

One is that packaging prevents the movement of moisture. When food is deep-frozen the moisture inside it is converted into tiny ice particles inside the cells of the food. It is absolutely essential that none of this frozen moisture should be allowed to escape from the food and conversely that no moisture from outside the food be allowed into it. You may notice that if you put unwrapped food into an ordinary refrigerator it soon becomes dry and unpalatable. But if you wrap it in, say, plastic film it will stay succulent and fresh much longer. The same thing applies, but to an even more important extent, in deep-frozen food, the packaging material forming a sort of wall through which the moisture cannot pass.

If unwrapped food were placed in the deep freeze, loss of internal moisture would render it dry, hard and uneatable. Conversely, since there is always some moisture even in the coldest air, it could take moisture from the air in the form of frost which would have just as harmful an effect on some foods.

The effect produced on the food itself by either of these two conditions is known as 'freezer-burn', which manifests itself in discolouration of the food, or spottiness, possibly mould; no amount of special cooking will take away the spongy tastelessness that results.

Another basic reason for packaging is to prevent contamination of one food by another. A deep-freeze cabinet is bound to contain a fairly wide variety of foods with different flavours and in many cases one might absorb the taste of another with disastrous effects. The case which the manufacturers delight in quoting is that of curry-flavoured ice cream.

Packaging therefore is a form of insulating food against air and other influences and must have these qualities:

It must be proof against moisture and air.

It must be greaseproof, so that grease will not work its way through before the contents are frozen solid.

It must have no smell or taste of its own and must be capable of preventing tastes or smells from passing through it.

It must be thin enough not to take up unnecessary space.

It must be tough enough to withstand a certain amount of knocking about to which it might be subjected in the freezer, yet flexible enough to wrap fairly tightly round different sorts of food.

It must be able to withstand the low temperatures to which it will be subjected in the deep freezer.

Certain plastics might almost have been invented especially for deep-freezing and the one that seems to have all the necessary virtues is a polythene (of about 250 gauge) which is specially made to withstand low temperatures. Closely following it is aluminium foil, which is not quite as tough. Add to these mutton cloth for guarding against breakage, non-stick paper for keeping some items separated from their own kind (chops, slices of cake) and an adhesive tape for sealing at low temperatures and you have the basic materials for freezer packaging.

The variety of these manufactured materials is almost infinite and they can be relied on to do their job, whereas any home-made packaging should be very well tested out before you commit valuable food to it for any length of freezer storage.

Polythene

Polythene comes in sheet wrapping form, bags, self-cling film or manufactured boxes. Thick polythene bags are specially made for the freezer. The bags can be moulded tightly around irregular-shaped foods such as, say, chickens, excluding most of the air from the packaged item. Plastic coated ties may be used to keep out the air, although some people prefer to heat-seal the bags by the use of the edge of an electric iron set to its coolest, or an old pair of hair curling tongs. They can also be sealed with special freezer sealing tape which will stay put even at the coldest temperatures.

Plastic containers

There is a very wide range of plastic containers on the market which have been designed for use in freezers and may have

Packaging - how and why continued

snap-on lids which do away with the need for tape-sealing since they are airtight. The initial cost of these is higher than some other packaging materials but they can be used over and over again for years and are particularly suitable for liquids and sauces.

Aluminium foil

This is easy to use and moulds well to the shape of the food. It is also useful for lining dishes in which food, such as stews or casseroles, may be frozen. When it is frozen the foil-wrapped food is removed from the dish and placed in a polythene or waxed paper overwrap to obviate any risk of the foil tearing.

Aluminium foil dishes

All sorts of shapes and sizes now come in heavy-gauge aluminium foil including pie plates, pie dishes, individual pie dishes and so on. These can be used a number of times, depending on what has been cooked in them and how easily they can be cleaned. They have the advantage that they can be placed in the oven with the thawed-out food still in them for reheating.

Overwrap for bags

Some freezer users put polythene bags into overwrapping of waxed or plastic-lined bags or boxes to prevent damage to the polythene bag.

Waxed boxes and cartons

Square, round and oblong waxed board containers are made in many sizes and are useful because they stack neatly, are easy to use, and may be used over and over again.

Glass and china containers

The temperatures in a freezer are too low for long-term storage of food in ordinary glass containers. Specially strengthened glass can be obtained but there is little need for this when polythene bags will serve as well. Again, many glass containers are round, the worst shape for economical storage, and most china dishes are too costly to have out of commission in the freezer for long periods.

A notable exception to the glass range is a special glass ceramic sold as Pyrosil, which could be used for freezing and can be taken straight from the freezer and placed in the oven. But at the price of this type of dish, using it for storage is expensive.

How much to package

Where deep freezing is concerned you might almost use the old adage 'Good things come in small bundles'. The smaller the quantity you freeze the quicker it will freeze and therefore the better it will keep. The same applies when it comes to thawing it out. But most important of all when considering how much to put in one packet is the question of accessibility.

It's all very well to make enough casserole for a dozen

or more servings, but you will find yourself in a bit of a fix if you freeze it all in one piece, then want only enough for four people. On the other hand if you package and freeze it in quantities for four, when it comes to taking it out you simply take out one package and return the rest to storage. Try cutting off a chunk of frozen casserole and see how you get on!

Put chops or steaks, for instance, between pieces of non-stick paper or foil singly or in pairs; the same goes for slices of cake, small cakes, buns and the like. Sandwiches should be packaged singly, both for convenience and to prevent mixing the taste of fillings; sliced bread may be rewrapped in half-loaves. No matter how much of a sauce you make pack it in $\frac{1}{2}$-pint lots and remember that a couple of pounds of green beans will still be as many if you pack them in four $\frac{1}{2}$-lb lots and life will be a lot easier.

Choose your shape

There is not much you can do about the shape of a joint of beef; it's there, it's solid and it has to stay that way when it is put into the deep-freeze. But semi-solids and liquids are another matter and you should always remember that these can be frozen into any shape and that the most economical shape for storage is one with straight lines. A little reflection will show for instance that you would lose far less space in the freezer by packaging mince in flat square slabs rather than in round balls. You can freeze liquids of all kinds into square shapes by what is called 'pre-forming', for instance putting a polythene bag into a cardboard sugar or cereal packet, filling it with liquid, then freezing it solid and discarding the packet. The variations on this theme are almost infinite.

When you need air

Packaging solid foods like meat, bread and so on is best done by excluding all the air possible. But in the case of liquids, or liquids with solids in them, like stews, some space must be left inside the packet because liquids expand when they freeze and they have to have room to do so. This is called 'headroom' or 'headspace' and a rough rule of thumb is to leave half an inch above each half pint. In some cases where rigid containers are used and there is not enough liquid to fill it to the required height, crumpled freezer paper is used to fill the space and thus eliminate most of the unwanted air.

What to put it in

A pattern emerges in deep-freezing in the packaging materials used for different sorts of food. For instance meat, poultry, game and the like are first wrapped in stockinette (to protect them from freezer burn in case the covering is torn) and then in polythene bags. Things like separated eggs, soft fruits, and fruit purée mostly go into waxed pots or cartons, whereas ordinary fruit goes into polythene bags; ice cream usually

Packaging - how and why continued

comes in cartons or polythene containers ready to go into the freezer. Fish is usually covered with foil and sealed inside a polythene bag, pies go on foil plates inside polythene bags, ready to be cooked or reheated when they are taken out of the polythene. Things like stock, sauce or soup are usually in waxed tubs or small polythene containers and bread and cakes in polythene bags.

The importance of sealing

Whatever the packaging, sealing is also essential, except in the case of some plastic containers with snap-on lids which are airtight. Ordinary household sealing tape is not good enough since at very low temperatures it loses its adhesive properties. A special tape may be bought for the purpose.

Polythene and heavy cellophane may be sealed with heat, remembering that unless you put some brown paper on both sides of the material the melted plastic may ruin your iron, or stick the plastic to whatever it is standing on.

Plastic covered wire ties may be used to tie a polythene bag tight after you have squeezed all the air out or sucked it out with a straw.

Varying the colours of polythene bags, wrapping, seals etc. will help you to identify what is in the package, because after it has been in the freezer for a short time it becomes a frosty white colour that prevents you from seeing what is inside. But labelling is a far safer way of knowing what is what, and it is advisable to put not only the contents but the date it was frozen and the date by which it should be eaten (see chapter on storage times).

Here are a few other general hints on packaging:

The more you can cool your food before packaging and freezing the better.

Do not put hot food into waxed or similar containers or you will melt the wax and spoil the container.

Do not let unfrozen food come into contact with frozen food inside the freezer.

Make sure that your labelling materials are proof against extreme cold and moisture.

Some pre-cooked foods should be placed on a foil or paper tray before packaging, and separated so that they will not get broken while frozen and can be easily separated for cooking.

Thickened liquids tend to become thicker on freezing so prepare sauces slightly thinner than usual and those containing egg yolks and cream should be reheated in a double saucepan to stop them from separating.

Seasonings taste stronger after freezing, so go easy.

Do not put sauces, soups etc. that contain wine or vinegar into aluminium containers without first lining the containers.

Skim as much fat as possible from stews, etc. before freezing, or cut off as much fat as possible from meat, since it tends to go rancid.

Unsalted butter freezes better than salted.

How long will it keep?

In theory, food will keep indefinitely providing that it is properly packaged and quick-frozen at the right temperature, then stored at a constant temperature of 0°F or less. In practice, however, it has been found that after a certain time while the food may still remain quite safe to eat, it will have lost some of its characteristics such as nutrient value, taste, texture or colour. These changes take place in different foods after different lengths of time in the deep-freeze therefore different foods have different storage times, or time limits within which they should be thawed out and eaten if they are to retain their original attractiveness.

The following table is a general guide to how long the various foods should remain in deep-freeze storage. Obviously these times cannot be entirely accurate since they depend to some extent on how well packaging has been done, how often the deep-freeze has been disturbed and numerous other factors, but they will show the maximum time for which it is advisable to leave them before they are eaten.

Meats

Beef
Joints	8 months
Portions (steaks, etc.)	8 months
Minced	3 months
Offals	3 months
Sausages, unseasoned	3 months
Sausages, seasoned	2 months

Veal
Joints	6-8 months
Portions (escalops, etc.)	6 months
Offals	2-3 months

Lamb and Mutton
Joints	6 months
Portions	6 months
Offals	3 months

Pork
Joints	6 months
Portions	6 months
Salted joints and portions	3 months
Sausages, unseasoned	3 months
Sausages, seasoned	2 months
Shop-bought sausages	4 weeks
Bacon, green	4-6 weeks
Bacon, smoked	1-2 months

How long will it keep? continued

Poultry and Game

Chickens	1 year
Duck	6 months
Turkey	6 months
Game birds	6-8 months
Hare and rabbit	6 months
Giblets	3 months

Fish

White fish	6-9 months
Other fish	3 months
Shell fish, commercially quick-frozen	3 months

Composite dishes

Those including beef, lamb, chickens and game	2 months
Including veal, pork, turkey, goose, duck, offals, etc.	1-2 months
Fish	2 months
Pâtés, terrines, etc	4-5 weeks

Pastry, Pies, etc

Meat pies, cooked	3 months
Meat pies, uncooked	3 months
Fruit pies	3-6 months
Pie cases, uncooked shortcrust	3 months
Pie cases, puff or flaky	4 months
Pie cases, cooked (all types)	6 months
Pizza, complete but uncooked	3 months
Pizza, cooked	2 months

Vegetables

Asparagus	9-12 months
Aubergine	1 year
Beans, broad or French	1 year
Beetroot	6-8 months
Broccoli	1 year
Brussels sprouts	6 months

Vegetables continued

Carrots (young)	6 month
Cauliflower	6-8 months
Corn on the cob	1 year
Mushrooms	1 year
Peas	1 year
Peppers	1 year
Potatoes, small new	1 year
Potato, chipped	3 months
Spinach	1 year
Tomato (as purée only)	1 year

Fruit

All fruit may be stored up to one year. Apples, pears, avocados and some other fruits are frozen only as purée.

Desserts

Many desserts may be deep-frozen, particularly those based on meringue, ice cream, whipped cream, fruit, etc. It is not recommended to leave them in the deep freeze for more than two months, although depending on their content some might be quite unharmed after 6 or 9 months.

Bread, Cakes and Pastry

Bread, bought or home-baked	1 month
Bread, partly-baked	4 months
Unrisen bread dough, plain	2 months
Unrisen bread dough, enriched	1 month
Risen bread doughs	3 weeks
Sandwiches, depending on filling	1-2 months
Biscuits, unbaked	6 months (it is hardly worth freezing baked biscuits)
Cakes, with fat	6 months
Cakes, without fat	3 months
Cakes, with icing and/or filling	2 months
Pastry, general, baked	6 months
Pastry, general, unbaked	3 months

How long will it keep? continued

Miscellaneous	
Butter, unsalted	6 months
Butter, salted	3 months
Cream with more than 40% butterfat	3 months
Ice cream	3 months
Eggs, separated	9 months
Soups	4 months
Stock	4 months
Sauces	3-4 months, depending on seasoning
Herbs	3-6 months

Thawing it out

Most foods that have been stored in the deep-freeze must be thawed out properly before they are cooked or reheated and if we have emphasised the importance of sticking to the rules in preparing, packaging and freezing food, it is equally important to ensure that the rules are observed in thawing out. There are some foods that do not need thawing out, but they are the minority and so we can clear the decks of them before discussing those to which thawing applies.

Most deep-frozen vegetables not only do not need to be thawed but in fact retain their flavour, nutrients and appearance much better if cooked from frozen. Generally this will consist in putting them straight into boiling salted water for the time required to cook them, but they may also be steamed, cooked in butter, sautéed or cooked in other ways. The main exceptions are asparagus, broccoli, corn on the cob, beetroot, mushrooms and peppers. The first two should be partly thawed so that they will cook more evenly, the others fully thawed (for several hours or overnight).

In the case of most pre-cooked foods, complete thawing out is preferable. Some items such as casseroles, stews, dishes with pastry that have been completely pre-cooked, uncooked pastry such as pizza, bread for toasting, can be used straight away without being thawed but in some cases the flavour will not be as good.

There the non-thawing story ends. If you try to cook other foods from frozen, they will react by being tough, lumpy, tasteless, or in some cases such as chicken, possibly dangerous. They should be allowed to thaw gently in their packing until they have returned to the same state in which they were when you suspended their animation, so to speak, by quick-freezing. In this way the moisture which was turned into tiny ice particles in them will return to water; attempts to speed up the thawing process by immersing the food in hot water or blowing warm air on it will certainly not improve the quality of the food and if it is uncooked may well make it tough and unpalatable.

Thus the best method of thawing out is to take the food from the deep freeze, unpack it into the smallest pieces that are still wrapped and place it in the refrigerator on the day before you want to use it. Thawing times naturally vary with the food concerned and the size of its package but the following guide should prove useful (and remember that thawed-out food should be cooked as soon as possible after it is thawed):

Meats
(uncooked per 1 lb weight)

Beef, veal and lamb should be thawed out still wrapped for at least 5-6 hours in refrigerator or $2\frac{1}{2}$-3 hours at room temperature.

Pork must be thawed out in the refrigerator, same time as other meats.

Minced meats, wrapped, 10 hours in refrigerator, 2 hours at room temperature.

Thawing it out continued

Offals, unwrapped, 9 hours in refrigerator, 2 hours at room temperature (except tripe and tongue, 3 hours longer in the refrigerator).
Cooked meats, wrapped, 9 hours in refrigerator, 5 hours at room temperature.

Poultry

Chickens, small, 24 hours either in refrigerator or at room temperature.
Chickens, medium and large, 24-36 hours.
Ducks, 24-36 hours either in refrigerator or at room temperature.
Geese and turkeys, small, 36 hours.
Geese and turkeys, large, 2-3 days.
Jointed poultry may be unwrapped and separated and should thaw in 6-9 hours according to the size of the portions.

Fish

Uncooked fish should thaw out in the refrigerator for at least 12 hours or overnight.
Cooked fish at least 2 hours, according to recipe.

Other foods

Bread, scones, croissants etc. overnight in the refrigerator or at least 3 hours at room temperature.
Pies, etc. in which the pastry is uncooked, can go straight into the oven but may also be thawed out 8 hours in refrigerator.
Cooked pastry which is not to be reheated, 8 hours in refrigerator.
Dessert dishes, depending on recipe, 5 or 6 hours in the refrigerator.
Iced cakes, 4 hours in refrigerator.
Eggs, 5-6 hours in refrigerator.
Soups, sauces etc. which are creamed should be heated in a double saucepan; others may be heated ordinarily.
Frozen fruits which are to be reheated may go straight into the pan. Those which are to be eaten now should thaw out 3–4 hours at room temperature.
Butter, 1 hour per lb at room temperature.
Sandwiches, 3 hours at room temperature.

The basic rules

Before we go on to the actual business of preparing, packaging and deep-freezing various uncooked and cooked foods, here again are the ten basic tenets of successful deep-freezing:

1. Use only the best quality and freshest foods and make sure all your equipment is kept scrupulously clean.
2. Make sure that prepared foods are suitably treated, remembering that thickened liquids become thicker and seasonings more emphatic after freezing.
3. Make sure your food is correctly wrapped in the right-sized portions and properly sealed.
4. Be certain that the temperature in the quick-freeze part of your deep-freeze is sufficiently low, i.e. $-12°F$.
5. Cool the food as much as possible so that when you put it into the quick-freeze section of your freezer it will freeze as quickly as possible and thus prevent the formation of large ice crystals inside the food.
6. Make sure the temperature of the storage part of your deep-freeze is below $0°F$.
7. Although it may sound trite, make sure your freezer cannot be switched off accidentally and that it is in a place where there is a good circulation of air.
8. Never store frozen food longer than the recommended time. To ensure this, label and date your food and use it in strict rotation.
9. Thaw out frozen food according to the rules. Trying to cut corners might result in a lot of waste.
10. The most important of all— **never re-freeze anything that has been fully thawed out.**

Freezing uncooked food

Food	To prepare	Packaging
Meat Joints of beef, lamb, and veal	Bone, roll and tie into shape (no skewers, please)	Double thickness of kitchen foil with stockinette overwrap
Pork	(As above but with extra care to make sure the meat is freshly slaughtered)	
Steaks and chops	Remove surplus fat from steaks; wipe over and separate with foil or polythene	Double foil or freezer paper with polythene bag or overwrap
Veal escalopes	Egg and crumb as for cooking, then pack with foil dividers	Wrap tightly in double foil, then seal in polythene bag
Offals	Soak in cold water, drain and rinse, removing fat, etc.	Polythene bags, with tie or heat seals.
Mince	Must be fresh and unseasoned.	Wrap tightly in sealed polythene.
Sausages	Must be fresh and unseasoned.	Wrap tightly in sealed polythene.
Poultry Whole chickens, ducks, geese, turkeys.	Pluck, dress, wipe inside and out, wrap and seal giblets and place inside. Cover bones with foil.	Seal in freezer bag with as much air as possible extracted.
Pieces of chicken, etc.	Wrap separately and tightly. May be egg-and-crumbed first.	Seal into separate polythene bags.
Game Game birds	As for poultry, but the game must be hung for the required time **before** freezing.	
Other game	Bleed, hang, clean and joint.	Wrap individually in polythene with stockinette overwrap.

Food	To prepare	Packaging
Fish		
Whole fish, salmon etc.	Scale, clean and gut, wash, drain and dry. Freeze unwrapped until solid, dip in water and freeze again, until covered with ice.	Overwrap in heavy polythene; use board for support if needed.
Fish fillets or steaks.	Separate with foil or polythene.	Wrap tightly in polythene and seal.

Vegetables
(Most vegetables should be blanched before they are frozen. This is generally done in a close-mesh wire basket in a large pan of boiling water, sufficient to cover the vegetables. All vegetables not being blanched should be washed.)

Asparagus	Trim, blanch 3–5 mins.	Pack in waxed paper or sealed box.
Aubergines	Peel and slice, blanch 4 mins.	Pack in plastic or waxed cartons.
Beans, broad	Shell, blanch $1\frac{1}{2}$ mins.	Polythene bags.
Beans, French	Top and tail, blanch 1–2 mins. if sliced, 2–3 mins. if left whole.	Polythene bags.
Beans, runner	Top and tail, blanch 2 mins. if sliced, $2\frac{1}{2}$ mins. whole.	Polythene bags.
Beetroot	Cook, peel, slice small.	Sealed boxes.
Broccoli	Cut into 6-inch sprigs, blanch 2–3 mins.	Polythene bags.
Brussels sprouts	Pick over, blanch 3–5 mins.	Polythene bags.

Freezing uncooked food continued

Food	To prepare	Packaging
Carrots (young)	Scrape, blanch 2–3 mins.	Polythene bags.
Cauliflower	Cut into florets, blanch $2\frac{1}{2}$ mins.	Lined boxes.
Corn on the cob	Remove leaves and threads, trim stem, blanch 6–10 mins.	Polythene bags.
Mushrooms	Wash in lemon juice and water, trim, blanch 2 mins.	Lined boxes.
Peas	Shell, blanch $1\frac{1}{2}$ mins.	Polythene bags.
Peppers	Discard stem and seeds, chop flesh, blanch 1–2 mins.	Plastic or waxed cartons.
Potatoes, new small	Scrape, wash, dry, blanch 3 mins.	Polythene bags.
Potatoes, chipped	Fry until half-cooked, drain and cool.	Polythene bags.
Spinach	Wash and dry, blanch 1 min.	Polythene bags.
Tomatoes (as purée only)	Scald and skin, then sieve pulp.	Plastic or waxed cartons.
Fruit Apples (as purée)	Peel, core, slice, stew and sieve.	Plastic or wax cartons.
Apricots	Peel, halve and stone.	Cover with syrup (if for jam or chutney pack dry in bags).
Avocado pears	Purée (raw)	Polythene bags.

Food	To prepare	Packaging
Fruit		
Blackberries	Pick over, wash, dry.	Cover with syrup or pack dry in polythene bags (with or without sugar).
Cherries	Stalk, wash; stone (optional).	Cover with syrup or pack in polythene bags.
Cranberries	Wash and dry.	Polythene bags.
Currants, red or black, etc.	Pick over, wash, dry.	Cover with syrup or pack dry in polythene bags.
Damsons	Halve and stone.	Cover with syrup.
Gooseberries	Top and tail, wash and dry.	Polythene bags.
Grapes	Peel, halve and pip.	Cover with syrup.
Peaches	Peel, halve and stone; slice (optional).	Cover with syrup with ascorbic acid added (optional).
Pears (purée only)	Peel, slice, core, stew and sieve.	Plastic or waxed cartons.
Plums and greengages	Wash, halve; stone (optional).	Cover with syrup.
Raspberries and loganberries	Pick over; wash in cold water, if necessary dry.	Plastic or waxed containers (with or without sugar).
Rhubarb	Wash and slice.	Cover with syrup or pack in plastic or waxed containers (with or without sugar).

Freezing uncooked food continued

Food	To prepare	Packaging
Strawberries (sliced or whole)	Hull, wash in cold water if necessary and dry.	Cover with syrup or pack in waxed or plastic containers with or without sugar.

Note: Ascorbic acid crystals should be used at the rate of $\frac{1}{4}$ tsp per lb. of finished fruit, that is fruit plus sugar. Use it also for preventing discolouration in other fruits. Do not try to freeze over- or under-ripe fruit and remember that it is not practicable to blanch fruits if they are to be used raw after defrosting.

Miscellaneous		
Butter	Use unsalted butter.	Wrap tightly in foil with polythene overwrap.
Cream	Only that with more than 40% butterfat.	In waxed or plastic tubs and sealed.
Eggs	Separated.	In waxed or plastic tubs.
Herbs	Freshly chopped.	In small polythene bags, sealed.

Cordon Bleu freezing recipes

We have shown you how you may freeze, store and then use food which has not been cooked, but the real delights of a deep-freeze are yet to come. Many of the famous Cordon Bleu recipes can be adapted with very little trouble to be suitable for deep-freezing either after they have been completely cooked, or with practically nothing more to be done outside of reheating when they are thawed out later for a special occasion or to accommodate the surprise guests, or simply just as part of a well-planned cook-ahead programme. Remember, when you are cooking these meals, that freezing will accentuate most of the usual seasonings and salt, so go easy on them and adjust when you are reheating. The quantities in the following recipes also have been left in the normal 4-person range but if you want to increase this, multiply the ingredients.

Consommé madrilène

3 pints chicken stock
¾ lb lean shin of beef (finely shredded)
1 lb ripe tomatoes (skinned, seeds removed and sliced)
2½ fl oz sherry, or Madeira
whites and shells (wiped and crushed) of 2 eggs
1 rounded dessertspoon tomato purée (optional)
extra sherry (optional)
1 extra tomato (skinned, seeds removed, and cut in fine shreds)—to garnish

Method

Put the stock into a thick enamel, or tin-lined, pan. Add the beef, tomatoes and the sherry (or Madeira). Using a metal whisk, whip egg whites to a light froth and add to liquid with the crushed shells; whisk backwards, over moderate heat, until the liquid is at boiling point. Then stop whisking and allow the soup to come up to the top of the pan. Draw pan aside, then replace on heat and boil up carefully once more, taking care not to break the 'filter' (froth of egg whites) which will form on top. Draw pan aside and leave for 40 minutes on very low heat to extract all the flavour from the meat.

Place a scalded cloth over a bowl and pour soup through, at first keeping the 'filter' back with a spoon and then, at the end, sliding it out on to the cloth. Pour the soup again through the cloth and the 'filter'. Consommé should now be clear; if not, pour back through the cloth. If wished, add tomato pureé to give a little colour and additional flavour. The best way to do this is to put the purée in a bowl and pour on the soup.

Cool rapidly, pour into a rigid plastic container and freeze rapidly. To use, hold container under cold water until the frozen soup can be turned out into a saucepan. Heat gently, stirring occasionally, and bring slowly to the boil. Adjust seasoning, add extra sherry if wished and add the garnish.

Consommé madrilène, a clear soup, is made on a base of chicken stock. Each serving is garnished with thin strips of tomato

Potage bonne femme

5 leeks
4 small potatoes
2½ oz butter
salt
pepper (ground from mill)
¾ pint milk
¾ pint water
1 teaspoon chopped parsley
1 extra leek, for garnish

For liaison
2 egg yolks
¼ pint single cream

If you prefer to serve a home-made soup as a first course, try leek and potato soup. It is easy to make and the addition of egg yolks and cream turns it into a party dish.

Croûtons

Tiny cubes of stale white bread are fried in shallow or deep fat until golden-brown. After draining them on absorbent paper, salt lightly.

Method

Wash and trim leeks very carefully as they can be gritty. Make a deep cross-cut through the leaves and wash under a fast-running tap. Slice the white parts only of the leeks. Peel and slice potatoes.

Melt the butter in a large pan, add the vegetables and seasoning and stir over a gentle heat until almost soft. When vegetables are well impregnated with the butter and have started to cook, cover them with a buttered paper and lid (this is known as sweating); there is no need to stir them all the time. Cook for at least 10 minutes.

Pour on the milk and water and stir until boiling. Draw pan half off heat and half cover the pan with a lid; leave to simmer for 15 minutes.

Pass through a Mouli sieve or mix to a purée in an electric blender.

Cool rapidly, pour into a rigid plastic container and freeze.

To use, turn the soup out into a saucepan and heat gently, stirring occasionally. Bring to the boil. Meanwhile, cut the extra leek into fine shreds, put into cold water and boil for 2 minutes. Drain and dry.

Work the egg yolks with the cream in a bowl, add to the soup as a liaison and gently reheat, without boiling; stir over gentle heat until the soup thickens. Adjust seasoning and garnish with chopped parsley and shredded leek. Serve fried croûtons separately.

Potage bonne femme – a leek and potato soup, with croûtons served separately

Mulligatawny

1 lb lean mutton, or lamb (a piece of double scrag is suitable)
2 onions
1 carrot
1 small cooking apple
1 large tablespoon dripping, or butter
1 dessertspoon curry powder
1 rounded teaspoon curry paste
1 rounded tablespoon plain flour
$2\frac{1}{2}$ pints cold water
few drops of lemon juice
$\frac{1}{4}$ pint milk

For liaison (optional)
little arrowroot
1–2 tablespoons water

This soup can be made from any kind of meat or trimmings of meat.

A good stock can be used instead of water in mulligatawny; in this case the meat will not be needed. For a richer soup, add a little cream to the milk.

A curry paste adds to the flavour; it is more spicy and blander than a curry powder and the two mix well together. This mixture, though not essential, is often used in curries.

Method

Soak meat for 1 hour in salted water. Slice vegetables and apple. Wipe and dry meat. Melt the fat in a pan and brown the meat lightly in it. Remove meat, add vegetables and apple; cook for 3–4 minutes. Add curry powder and paste.

Watchpoint The curry powder is first fried for 2–3 seconds to ensure that it is cooked.

After 2 minutes stir in the flour and pour on the water. Bring to the boil and add the meat. Cover and cook gently for $1-1\frac{1}{2}$ hours. Then take out the meat and sieve or blend liquid and vegetables. If using an electric blender, add some of the meat. Add the lemon juice. Cool rapidly, pour into a rigid plastic container and freeze.

To use, loosen the block of frozen soup from its container by holding in cold water. Turn into a saucepan and heat gently, stirring occasionally. When thawed, add the milk and bring to the boil. Thicken if necessary with a little arrowroot slaked in cold water.

Chicken broth

2 pints strong chicken stock
3 tablespoons carrot (finely diced, yellow core discarded)
2 tablespoons finely chopped onion
2 tablespoons long grain rice
salt and pepper
3 tablespoons double cream
1 dessertspoon chopped parsley

Method
Remove all fat from stock, and put stock, vegetables and rice in a pan, season, cover and simmer for 30–40 minutes. Cool rapidly, pour into a rigid plastic container and freeze. To use, turn out into a saucepan and heat gently, stirring occasionally. Bring to the boil and stir in the cream. Adjust seasoning, sprinkle with parsley and serve.

Cockie-leekie soup

1 large boiling fowl
4–6 pints cold water
6 leeks
2 tablespoons rice
salt and pepper
1 tablespoon chopped parsley

This traditional Scottish soup is made with an old cock bird, hence its name.

Method
Place the boiling fowl in a large pot. Add cold water to cover the bird, salt lightly, cover and simmer for 2 hours.
 Make a deep cross-cut in the leeks, wash well and cut into slices. Skim fat off chicken liquid, add leeks and rice and cook gently for a further $1\frac{1}{2}$ hours. Set aside the bird and cool the soup. When cold, remove any fat from the surface. Take a slice or two of meat from the leg of the fowl and cut into fine shreds; add to the soup. Pour into a rigid plastic container and freeze.
 To use, turn out into a saucepan and heat gently, stirring from time to time. Bring to the boil, adjust seasoning and sprinkle with parsley.

Vegetable bortsch

beetroot
onions
carrots
celery
1 parsnip
salt and pepper
stock (preferably ham), or water
cabbage (coarsely shredded)
garlic (chopped, or crushed)—to taste
tomatoes
sugar
little tomato purée
fresh parsley (chopped)

For liaison
little flour (optional)
soured cream

5-inch diameter pudding basin (sufficient for 3 pints liquid), or small mixing bowl

Quantities of vegetables should be used in the following proportions: half beetroot and of remaining half, one-third onion, one-third carrot and the last third equally divided between celery and parsnip.

Method

Cut beetroot, onions, carrots, celery and parsnip into matchsticks and pack into the basin or bowl to fill it.

Lightly season stock or water and bring to the boil. Turn the bowl of vegetables into the pan, cover and simmer for about 20–30 minutes. Coarsely shred enough cabbage to fill the bowl, add this with the garlic to taste. Continue to simmer gently, uncovered, for a further 20 minutes.

Skin sufficient tomatoes to half-fill the bowl, squeeze to remove seeds, then chop flesh very coarsely. Add to soup, season well with salt and sugar and add a little tomato purée to sharpen the flavour. Simmer for a further 10 minutes. Cool rapidly, pour into a rigid plastic container and freeze.

To use, turn out into a saucepan and heat gently. Bring to the boil and add a handful of chopped parsley.

The soup can be thickened lightly with a little flour mixed with a small quantity of soured cream. Otherwise serve a bowl of soured cream separately.

Watchpoint Bortsch should be slightly piquant in flavour and not sweet. Add salt and sugar until this is reached. The soup should be a thick broth of vegetables but not too solid. Dilute if necessary with additional stock.

A colourful show of vegetables for making bortsch, a soup that's slightly piquant in taste

Cream of asparagus soup

2 bundles of sprue, or 1 bundle of asparagus
1½ pints veal, or chicken, stock
1 small onion (finely chopped)
1 oz butter
¾ oz plain flour
salt and pepper
extra asparagus tips, for garnish (optional)

For liaison
2 egg yolks
1 small carton (about 2½ fl oz) double cream

Method

Wash and trim the tied bundle of sprue or asparagus, then cut prepared stalks into 1-inch pieces and put these in a pan with the stock and onion. Cover pan and simmer until asparagus is tender. Rub the soup through a nylon sieve or work in a blender.

Rinse out the pan. Make a roux with the butter and flour, then add the sieved or blended liquid and season. Bring this to the boil, then simmer it for 2–3 minutes. Cool rapidly, pour into a rigid plastic container and freeze.

To use, turn out into a saucepan, heat gently and stir from time to time. Add the egg yolk and cream liaison and reheat carefully, without boiling. Adjust seasoning, and if possible add a few asparagus tips as a garnish.

Chilled prawn bisque
(Bisque de crevettes glacée)

6 oz shelled prawns (chopped)
1 onion (finely chopped)
1 oz butter
2 lb tomatoes, or 1 medium-size can (1 lb 14 oz) tomatoes
3 caps of canned pimiento (chopped)
1 dessertspoon tomato purée
2–2½ pints chicken stock
arrowroot
¼ pint double cream

This soup can also be served hot.

Method

Cook the onion in the butter until softened, then add the tomatoes (skinned, cut in half and squeezed to remove the seeds). Cover the pan and slowly cook the vegetables to a pulp. Add the pimiento, tomato purée and stock. Simmer for 10–15 minutes. Then add the prawns and work in an electric blender.

Note: if a blender is not used, pass the vegetables and liquid through a fine sieve and add the prawns, finely chopped, after sieving.

Chill rapidly, pour into a rigid plastic container and freeze. To use, place the container of frozen soup in the refrigerator for 6–8 hours, to thaw; thicken slightly with arrowroot if necessary. Whip the cream and stir it into the soup just before serving.

Iced curry soup

1 oz butter
4 shallots, or 1 medium-size onion (finely chopped)
1 tablespoon curry paste
1 oz plain flour
1¾ pints chicken, or well-flavoured vegetable, stock
strip of lemon rind
1 bayleaf
¼ pint boiling water
1 tablespoon ground almonds
1 tablespoon desiccated coconut
1 dessertspoon arrowroot
1 tablespoon cold stock, or water

For cream topping
1 glass port
1 teaspoon curry powder
1 dessertspoon apricot jam, or purée of fresh, or dried, apricots
4 tablespoons double cream

The liquid from the wine, curry and jam mixture is stirred into whipped cream to make the topping for iced curry soup

Method

Melt three-quarters of the butter, add the shallot (or onion) and cook it slowly until just turning colour, then add the curry paste and a dusting of the flour; fry gently for 4–5 minutes. Stir in the remainder of the butter and when it has melted blend in the rest of the flour and the stock; bring to the boil. Add the lemon rind and bayleaf and simmer for 20 minutes. Strain the liquid and return it to the rinsed pan; continue simmering for 10–15 minutes.

Meanwhile pour the boiling water over the almonds and coconut and leave them to soak for 30 minutes, then squeeze mixture in a piece of muslin and add the 'milk' obtained to the soup. Mix the arrowroot with the tablespoon of cold stock (or water), add it to the pan and reboil. Strain soup again, allow it to cool. Pour into a rigid plastic container and freeze rapidly. To use, thaw in the refrigerator for 6–8 hours.

To make the cream topping: mix the port and curry powder together and simmer until reduced to half quantity. Leave this until cold, then mix in the jam (or purée) and squeeze the mixture in a piece of muslin; reserve the liquid. Lightly whip the cream and stir in the 'essence' from the wine and curry mixture. Serve the soup with a spoonful of this cream in each soup cup.

Iced curry soup, served in individual cups with a spoonful of cream topping in each

Coquilles St. Jacques armoricaine

5–6 good-size scallops
4–6 peppercorns
squeeze of lemon juice
1 bayleaf
1 medium-large carrot (finely diced)
2–3 sticks of celery (finely diced)
1 large, or 2 small, leeks (finely sliced)
1 oz butter
2–3 brussels sprouts (finely sliced)—optional
2 tablespoons white wine, or water, or 1–2 tomatoes (scalded, skinned, seeds removed)
1–2 tablespoons grated Cheddar, or Gruyère, cheese

For cream sauce
½ oz butter
½ oz flour
7½ fl oz creamy milk
salt and pepper

4–5 scallop shells

Method

Wash and clean the scallops and put them into a pan. Cover with cold water, add the peppercorns, lemon juice and bayleaf and bring to the boil; poach for 5–7 minutes.

Set the oven at 350°F or Mark 4. To prepare the mirepoix: put carrot, celery and leek in a flameproof casserole with butter; cover and cook on a low heat for 3–4 minutes. Add the brussels sprouts (if using) and the white wine (or water, or tomatoes). Cover and put in oven for 5–6 minutes.

Put a spoonful of the mirepoix into each scallop shell. Drain the scallops, slice them into rounds and lay these on top of the mirepoix. Prepare the sauce, adding any juice from the mirepoix to it. Season the sauce and spoon a little over the contents of each shell. Place the shells in the freezer uncovered and freeze rapidly. When solid, wrap in polythene film or bags and seal.

To use straight from the freezer, remove the packaging, place the shells on a baking tray and cover loosely with foil. Place in the oven, pre-set at 425°F or Mark 7 and cook for 1 hour. Remove the foil, sprinkle with cheese, increase the oven temperature to 450°F or Mark 8 and return the shells to the oven for 10–15 minutes to brown.

Alternatively, remove the wrapping and thaw at refrigerator temperature for about 8 hours. Place the shells on a baking sheet and heat in the oven, pre-set at 425°F or Mark 7, for 15 minutes. Increase the heat and finish as above.

Avocado pear and tomato ice

1 can (14 oz) tomatoes
1 clove of garlic (crushed)
½ teaspoon salt
2 tablespoons sugar
pared rind and juice of ½ lemon
3 sprigs of mint
1 stick of celery (sliced)
1 onion (sliced)
Tabasco, and Worcestershire, sauce (to taste)
3 avocado pears

Ice-cream churn freezer

This dish serves 6 as a starter.

Method

Tip the canned tomatoes into a pan, add the garlic, salt, sugar, lemon rind and juice, celery and onion and stir until boiling. Press the tomatoes well, then add the sprigs of mint. Cover the pan and allow to simmer for 5–10 minutes. Remove the mint and rub the contents of the pan through a nylon sieve (leaving behind celery and onion). Allow to cool, add the Tabasco and Worcestershire sauces to taste. This mixture should be rather over-flavoured as freezing takes some out.

Chill the mixture really well then turn into a churn freezer and churn until firm. Turn into a rigid plastic container and seal.

For serving, slightly chill the avocados, cut in half and remove the stone. Place a good scoop of the ice in the cavity.

Country-style pâté

1 lb veal, or pork (minced)
8 oz pigs liver (minced)
4 oz pork fat (minced)
1 shallot (finely chopped)
1 large wineglass port wine
about ¼ of a small white loaf
3 eggs (beaten)
small pinch of allspice
1 teaspoon chopped marjoram
pinch of salt
6–8 rashers of streaky bacon

Medium-size loaf tin

Method

Put the veal and pigs liver, pork fat and shallot into a bowl. Pour the port over the bread and leave until thoroughly soaked; add this to the meats with the beaten eggs, allspice, herbs and salt. Work together in electric blender, or beat thoroughly. Line the loaf tin with the bacon rashers, fill with the mixture and press well down. Smooth the top, cover with foil or tie on a double sheet of greaseproof paper. Cook in a bain marie for 1¼–1½ hours at 350°F or Mark 4. The pâté is cooked when firm to the touch. Press the pâté in the tin with a light weight (about 2 lb) and leave until cold. Turn out and cut into slices. Separate each slice with freezer paper or waxed paper and reshape into a loaf shape. Wrap in a double thickness of foil and overwrap with polythene. Freeze rapidly.

To use, take out as many slices of pâté as are required and cover them loosely with fresh foil; return the remainder to the freezer in the original wrappings.

Thaw in the refrigerator for about 18 hours and bring to room temperature for serving.

Chicken and calves liver pâté

8 oz thin streaky bacon rashers (unsmoked)
4 oz chicken livers (sliced)
½ oz butter
1 clove of garlic (crushed)
1 teaspoon chopped thyme
2 tablespoons chopped parsley

For farce
2 lb calves, or lambs, liver (in the piece)
milk
8 oz pork fat (minced)
8 oz lean pork (minced)
2 shallots (finely chopped)
¼ pint double cream
2 eggs
salt
pepper (ground from mill)
1 small wineglass brandy, or sherry
luting paste

This rich pâté is suitable for serving as a first course or as a light lunch dish.

Luting paste is a flour and water mixture of a consistency similar to that of scone dough. To seal a casserole or terrine, put 3–4 oz flour into a bowl and mix quickly with cold water to a firm dough (4 oz flour will take ⅛ pint water).

Method

Remove rind from the bacon, line a terrine with rashers. Sprinkle with a little of the brandy or sherry and grind over a little pepper from the mill. Set aside.

Remove any ducts and soak calves or lambs liver in milk for 2 hours. Then rinse and dry thoroughly. Cut in pieces and pass through a mincer. Mix with the minced fat, pork and the shallots. If possible work for a few seconds in an electric blender for additional smoothness. Mix in the cream, beaten eggs and rest of brandy or sherry. Season well.

Remove any ducts or veins from the chicken livers and slice. Sauté in butter for 2–3 minutes, add garlic and herbs and mix well.

Put half the farce into the terrine and scatter the liver mixture on the top. Cover with rest of the farce and put any remaining bacon rashers on the top. Cover with lid and seal with luting paste. Cook in a bain marie in the oven for 1–2 hours at 325°F or Mark 3, until firm to the touch. Remove lid, press, using about a 2 lb weight, and leave to cool. Turn out, wrap in foil and polythene and freeze. To use, unwrap the pâté, return the frozen shape to the terrine in which it was originally cooked and thaw in the refrigerator for 18–24 hours. Bring to room temperature for serving.

Cod's roe pâté

12 oz smoked cod's roe (in the piece), or an 8 oz jar
1 teaspoon onion juice (from grated onion)
¼ pint olive oil
1 cup fresh white breadcrumbs, or 3-4 slices of bread
1 packet Demi-Sel cheese
lemon, or tomato, juice (to taste)
pepper

To garnish
black olives
lemon quarters

Method

Scrape the roe from the skin and put it in a bowl with the onion juice. Pour the oil over the breadcrumbs and leave to soak for 5 minutes (if using slices of bread, remove the crust, put bread in a dish and sprinkle with the oil). Pound or beat the cod's roe with the Demi-Sel cheese until quite smooth, then work in the breadcrumbs and oil, a little at a time. Finish with lemon (or tomato) juice to taste and season with pepper. The mixture should be light and creamy.

Line 4 individual soufflé dishes with foil and divide the pâté between them. Freeze. When solid, remove the foil linings from the dishes, cover the pâté, over-wrap with polythene and seal.

To use, unwrap the frozen pâté portions and place them in the individual soufflé dishes. Thaw in the refrigerator for 12 hours. Garnish with black olives and quarters of lemon and serve with hot toast and unsalted butter.

1 *Scraping smoked cod's roe from the skin*
2 *Pounding the roe with Demi-Sel cheese*

Stuffed aubergines (aubergines farcies)

2 aubergines
4 lambs kidneys
1½ oz butter
2 medium-size onions (finely sliced)
1 dessertspoon plain flour
1 teaspoon tomato purée
¼ pint stock
1 clove of garlic (crushed with ½ teaspoon salt)
salt and pepper
1 bayleaf
½ lb tomatoes
2–3 tablespoons salad oil
1 tablespoon grated cheese (preferably Parmesan)
1 tablespoon fresh white breadcrumbs

Method

Split aubergines in two lengthways, score round edge and across, salt and leave for 30 minutes to dégorger.

Skin the kidneys and cut out cores, cut in half lengthways. Heat a small sauté pan, drop in half the butter and, when foaming, put in the kidneys. Brown quickly on all sides then remove from the pan and keep warm. Lower the heat, add remaining butter and the onion. Cook for 2–3 minutes then draw aside. Stir in the flour, tomato purée and stock and bring to the boil. Add the crushed garlic to the pan with pepper, bayleaf and the kidneys. Cover and simmer gently for about 20 minutes.

Wipe the aubergines dry and sauté rather slowly in 2–3 tablespoons of oil until soft.

Watchpoint Aubergines brown very quickly when sautéd. If the flesh is browned before being cooked right through, complete cooking in oven.

Skin the tomatoes, remove the seeds and roughly chop flesh. When the aubergines are tender, scoop out the pulp with a spoon, leaving the skins intact.

Remove bayleaf from the kidneys, add the tomatoes and aubergine pulp and simmer together for 2–3 minutes. Set the aubergine skins in a foil dish and fill with the mixture. Cover carefully with foil and overwrap with polythene. Freeze.

To use, remove the polythene over-wrap and loosen the foil. Place in the oven, pre-set at 400°F or Mark 6, and heat for about 40 minutes. Remove the foil, sprinkle with cheese and breadcrumbs and raise the temperature to 450°F or Mark 8. Return the dish to the oven until the topping is crisp and brown.

The stuffed aubergine halves

Dolmades

6 oz raw beef (minced)
1 small onion (finely chopped)
salt and pepper
5 tablespoons water
1 cabbage
flour
stock
1 bayleaf
½-¾ pint tomato sauce
parsley (chopped)

Method

Put the onion into a bowl with the meat and seasoning. Work well together, adding the water gradually until the mixture is well beaten and pliable.

Trim the cabbage, blanch it whole in boiling water for 2-3 minutes, drain well, then carefully detach the leaves, removing any hard stalk.

Put a small tablespoon of the meat mixture on each leaf, roll up like a parcel to form a sausage shape. Then roll very lightly in flour and arrange in criss-cross layers in a thick pan or flameproof casserole. Barely cover with stock, bring carefully to the boil, season, add a bayleaf and simmer for 45-50 minutes on the stove, or in the oven at 350°F or Mark 4.

Carefully lift the dolmades out of the liquor, drain well and cool quickly. Place them in a rigid container, separating each one with freezer paper, and freeze.

To use, unwrap the dolmades, place them in an ovenproof serving dish. Cover and heat in the oven at 350°F or Mark 4, for about 45 minutes. Meanwhile turn the tomato sauce into a saucepan and heat gently, stirring. When thawed, pour the sauce over the dolmades and heat them together for a further 15 minutes. Sprinkle well with chopped parsley before serving.

Tomato sauce

1 lb tomatoes, or 1¼ cups of canned tomatoes
½ oz butter
1 dessertspoon plain flour
7½ fl oz stock (made from a bouillon cube)
bouquet garni
pinch each of salt, pepper and sugar

Method

Melt the butter in a saucepan, stir in the flour and add the stock. Add the tomatoes (cut in half and seeds removed), bouquet garni, salt, pepper and sugar. Strain the juice from the tomato seeds and add it to the sauce.

Cover and simmer for 30 minutes, then strain and boil to reduce. Cool, turn into a rigid container and freeze. To use, see above.

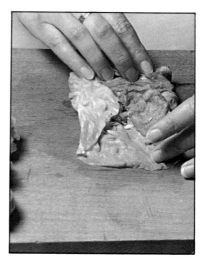

How to stuff cabbage leaves for dolmas. Take a softened leaf and spoon a little of the savoury filling into the centre. Fold leaf corners into the centre and roll up like a small parcel. Repeat this until all leaves are used. Then roll each of the finished leaves in a little flour

Quiche lorraine

For rich shortcrust pastry
6 oz plain flour
pinch of salt
3 oz butter, or margarine
1 oz shortening
2 tablespoons cold water

For filling
1 egg
1 egg yolk
1 rounded tablespoon grated cheese
salt and pepper
$\frac{1}{4}$ pint single cream, or milk
$\frac{1}{2}$ oz butter
2–3 rashers of streaky bacon (diced)
1 small onion (thinly sliced), or 12 spring onions

7-inch diameter flan ring

Hot or cold, this bacon tart is the most typical of dishes from the Lorraine region of France.

Method

Sift together the flour and salt, rub in the fats and bind to a smooth dough with the water. Set aside to chill.

When chilled, line the pastry on to the flan ring. Beat the egg and extra yolk in a bowl, add the cheese, seasoning and cream or milk. Melt the butter in a small pan, add the bacon and sliced onion, or whole spring onions, and cook slowly until just golden in colour. Then turn contents of the pan into the egg mixture, mix well and pour into the pastry case.

Bake for about 25–30 minutes in an oven at 375°F or Mark 5. Cool, place on a flat plate and freeze. When solid, wrap in polythene and seal.

To use, remove the polythene, cover loosely with foil, place on a baking sheet and reheat in the oven, pre-set at 400°F or Mark 6, for about 45 minutes; remove foil for the last 10 minutes.

Basic pizza dough

1 lb plain flour
1 teaspoon salt
1 oz fresh yeast
2 teaspoons sugar
about $\frac{1}{4}$ pint milk (warmed)
3-4 eggs (beaten)
4 oz butter (creamed)

Method

Sift the flour and salt into a warmed basin. Cream yeast and sugar and add to the warmed milk with the beaten eggs: add this liquid to the flour and beat thoroughly. Work the creamed butter into the dough. Cover and leave for 40 minutes to rise.
Note: for the best pizza, it is wise to use a flan ring to keep the dough in position. It has the added advantage of enabling you to cover the entire surface with topping without it running and sticking to your baking sheet.

To freeze the unbaked pizza dough, divide the above quantity into four, flour the dough lightly and pat it out with the palm of your hand on a floured baking sheet, to a round 8 inches in diameter. Freeze the rounds flat then wrap in polythene and seal.

To use, unwrap and cover with the required topping. Place the pizza in a cold oven, set it at 450°F or Mark 8 and bake for 30-35 minutes.

Pizza napolitana

$\frac{1}{4}$ quantity of basic dough

For topping
4-6 anchovy fillets
2 tablespoons milk
1 lb ripe tomatoes
1-2 tablespoons olive oil
1 small onion (finely chopped)
1 dessertspoon chopped marjoram, or basil
salt and pepper
4 oz Bel Paese, or Mozzarella, cheese (sliced)

8-inch diameter flan ring

Method

Flour the dough lightly and pat it out with the palm of your hand on floured baking sheet to a round 8 inches in diameter. Then place greased flan ring over it.

Split the anchovy fillets in two lengthways and soak them in the milk; set aside.

Scald and skin the tomatoes, cut away the hard core, squeeze gently to remove seeds, then slice. Heat the oil in a frying pan; add chopped onion and, after a few minutes, the sliced tomatoes. Draw pan aside and add the herbs; season well.

Set oven at 400°F or Mark 6.

Cover dough with tomato mixture, place cheese slices on this and arrange anchovies lattice-wise over the top. Prove pizza for 10-15 minutes, then bake in pre-set oven for 30-35 minutes. Lift off the flan ring and cool quickly. Freeze flat and wrap in foil or polythene.

To use, remove wrappings and bake in the oven, preheated to 400°F or Mark 6, for about 20 minutes.

Pizza Cordon Bleu

¼ quantity of basic dough

For topping
2 shallots (finely chopped)
1 wineglass white wine
1 lb scampi
4 oz mushrooms (chopped)
1 oz butter
¾ oz plain flour
1 clove of garlic (crushed with ½ teaspoon salt)
¼ pint chicken stock
1 teaspoon tomato purée
4 tomatoes
salt and pepper

8-inch diameter flan ring

Method

Simmer shallot in white wine until reduced to half the quantity. Add scampi and mushrooms and cook very slowly for 5 minutes; set pan aside.

Melt butter, add flour and when coloured add the garlic, stock and tomato purée, stir until boiling, then cook for 3–4 minutes. Scald tomatoes, skin, quarter, and remove seeds, cut flesh again into strips. Add scampi mixture to tomatoes and sauce. Season to taste.

Set oven at 400°F or Mark 6. Pat out the dough as before and cover with the topping; prove and bake in pre-set hot oven. Freeze and re-heat as for Pizza Napolitana.

Pizzas are popular party dishes. Choose from several different fillings: smoked haddock and mushroom (at back), ham (centre), napolitana (front left) and Cordon Bleu (front right)

Smoked haddock and mushroom pizza

¼ quantity of basic dough

For topping
1 lb smoked haddock
béchamel sauce (made with 1 oz flour, 1 oz butter, ¼ pint flavoured milk)
1 oz butter
1 shallot (finely chopped)
6 oz mushrooms (quartered)
salt and pepper

8-inch diameter flan ring

Method
Cover the smoked haddock with water, bring it slowly to the boil; cover, turn off heat and leave for 10 minutes.
 Meanwhile make béchamel sauce in the usual way.
 Remove skin and bones from the haddock and flake flesh carefully. Melt 1 oz butter, add shallot, cook for 2–3 minutes, then add quartered mushrooms and sauté briskly for 2–3 minutes. Add béchamel sauce and haddock; season to taste.
 Set oven at 400°F or Mark 6. Pat out the dough as before, cover with topping, prove and bake in pre-set hot oven.
 Freeze and re-heat as for Pizza Napolitana.

Ham pizza

¼ quantity of basic dough

For topping
2 oz butter
1 large Spanish onion (about ½ lb)— finely sliced
6 oz ham (shredded)
2 oz mortadella sausage (shredded)
2–3 tablespoons mango, or tomato, chutney

Method
Melt butter, add onion and cook slowly until very brown. Add shredded ham and mortadella, moisten with chutney.
 Set oven at 400°F or Mark 6. Pat out the last quarter of dough as before, cover with the topping, prove and bake in pre-set hot oven.
 Freeze and re-heat as for Pizza Napolitana.

Salmon mousse

¾ lb salmon steak

For court bouillon
¾ pint water
juice of ¼ lemon, or 1 wineglass white wine
½ teaspoon salt
3 peppercorns
bouquet garni

For béchamel sauce
¾ cup of milk
½ bayleaf
1 blade of mace
6 peppercorns
1 slice of onion
3 oz butter
1 oz plain flour
salt
2 tablespoons double cream (lightly whipped)
1 tablespoon medium sherry
2 drops of carmine colouring (optional)

To finish
½ pint aspic jelly (cool but still liquid)
¼ cucumber (thinly sliced)

6-inch diameter top (No. 2 size) soufflé dish

This is a very rich mousse and you serve about 1 tablespoon as a portion. A cucumber salad could be offered, too, or just Melba toast.

Method

First make the court bouillon, then put the salmon in a large pan, cover it with the hot court bouillon, bring this to the boil, then reduce the heat. Cover the pan and cook salmon very gently for 15 minutes; allow fish to cool in the liquid, then drain it on absorbent paper and remove all the skin and bone.
Watchpoint The court bouillon must just 'tremble' throughout the cooking time. If you want to cook the salmon in the oven, cover it with greaseproof paper and baste it frequently; allow 30 minutes at 325°F or Mark 4 (settings aren't comparable for this). Prepare béchamel sauce then turn on to a plate to cool.

Cream the remaining butter until soft and lightly whip the cream. Work the salmon in a bowl with a wooden spoon or the end of a rolling pin, or pound with a pestle in a mortar.
Watchpoint Pounding the salmon is important: this breaks down the fibres of the fish and it will then hold the sauce and butter without them curdling.

Add the cold béchamel sauce and butter to the salmon and taste for seasoning. Fold in the cream and sherry. Add the carmine if the salmon is a very pale colour.

Turn mousse into a foil lined soufflé dish, smooth the top with a palette knife and set in a cool place for about 10 minutes to firm. Pour over a thin layer of aspic jelly and when set arrange the cucumber on top, dipping each slice first in liquid aspic. When this garnish is set in position fill dish to the top with the remaining aspic.

Freeze until firm then remove the soufflé dish. Overwrap the foil with polythene and seal.

To use, remove the wrappings, place the mousse in the soufflé dish and thaw in the refrigerator for about 6 hours, or at room temperature for about 2 hours.

Seafood flan

½–¾ lb mixed shellfish (eg. prawns, scallops and mussels)
¼ lb button mushrooms (sliced)
½ oz butter
1 green pepper (blanched, seeds and core removed, and shredded)
6 oz quantity of flaky, or puff, pastry
1 egg (beaten)

For sauce
1 shallot (chopped)
1 oz butter
1 teaspoon curry powder
1 oz plain flour
scant ¾ pint milk
salt and pepper
2 tablespoons double cream

Deep 8-inch diameter pie plate, or 9-inch long pie dish

Method

First prepare filling: sauté mushrooms in ½ oz butter. Prepare the shellfish; if using scallops or mussels have them previously cooked and the prawns shelled or carefully thawed out. Mix shellfish with the green pepper and mushrooms.

To prepare sauce: soften shallot in the butter, add curry powder and, after 1 minute, stir in the flour. Pour on milk, blend, then stir until boiling. Season and finish with the cream. Pour sauce over the shellfish, mix together carefully and turn into the pie plate or dish, lined with foil. Roll out pastry to a thickness of ¼–½ inch and chill for 10–15 minutes. Set oven at 400°F or Mark 6.

When mixture is cold cover with the pastry; decorate it and brush lightly with beaten egg. Bake in pre-set hot oven for 25–30 minutes or until well browned.

Cool and freeze. Remove the pie plate or dish, cover with foil and overwrap with polythene.

To use, remove wrappings, return to the pie plate and thaw at room temperature for about 2 hours. Set the oven at 400°F or Mark 6.

Cover the flan loosely with foil and heat for 20–30 minutes; remove the foil for the last 10 minutes.

Sole Georgette

1 filleted sole (weighing 1½ lb)
4 large long-shaped potatoes
1 wineglass white wine
½ wineglass water
slice of onion
6 peppercorns
¾ oz butter
1 rounded tablespoon plain flour
salt and pepper
5 tablespoons top of milk
4 oz prawns (shelled)
grated cheese (to sprinkle)

Forcing bag and large rose pipe

Method

Set the oven at 375°F or Mark 5.
 Scrub the potatoes well, dry and roll them in salt. Bake until tender (about 1½–2 hours).
 Wash and dry the fillets; fold them over to the length of the potatoes. Place fillets in an ovenproof dish, pour over the wine and water, add the onion and peppercorns. Poach in a slow to moderate oven (325–350°F or Mark 3–4) for 10–12 minutes.
 Melt the butter in a saucepan, stir in the flour and strain on the liquid from the fish. Season. Blend and stir until boiling. Add the top of the milk, adust seasoning. Add 1–2 tablespoons of the sauce to the prawns to bind them.
 When potatoes are soft, cut off the tops lengthways, scoop out the pulp, divide the prawn mixture evenly between the potatoes and place inside the skins. Lay a fillet of sole on top and coat with the rest of the sauce. Make a purée of the

1 *Scooping out cooked potato from the skin to make container for fish; alongside are prawns mixed with a little of the sauce*
2 *Piping potato purée round edge of the potato skin after it has been filled*

Sole Georgette continued

scooped-out potato and pipe round the edge of each potato skin.

Cool and freeze. Wrap and seal individually in polythene or lay in a single layer in a large, rigid container.

To use, remove wrappings and place the potato cases on a baking sheet. Sprinkle with grated cheese and place in the pre-heated oven at 400°F or Mark 6 for about 30 minutes.

Fish croquettes vert-pré
with tartare sauce

1 lb fresh haddock fillet
salt and pepper
6 peppercorns
squeeze of lemon juice
2 tablespoons water
½ lb potatoes
1 bunch of watercress
1 oz butter
1 egg

For coating croquettes
2 tablespoons seasoned flour
1 egg (beaten)
dried white breadcrumbs

Deep fat bath

Method

Wash the haddock, then place in a lightly buttered ovenproof dish. Season with salt, add the peppercorns, lemon juice and water. Cover with a buttered paper and poach fish in a moderate oven, pre-set at 350°F or Mark 4, for 12–15 minutes. Drain fish, then flake and crush with a fork.

Peel and quarter the potatoes; wash the watercress well; cook both together in a pan of boiling, salted water until the potatoes are just tender. Drain well, dry both over gentle heat, then pass through a sieve.

Mix the fish and potato and watercress purée together, add the butter, season to taste, add egg and beat well. Form mixture into croquettes, using a palette knife to shape the ends. Roll the croquettes in seasoned flour, coat with beaten egg and white crumbs. Fry croquettes in deep fat until golden-brown (about 3 minutes) then drain on absorbent paper.

Cool and freeze. Wrap in foil and polythene and seal. To use, remove wrappings and place on a baking sheet. Re-heat in the oven, pre-heated to 350°F or Mark 4, for about 30 minutes. Serve with tartare sauce.

Quenelles de brochet

1 lb turbot, or halibut, steak
¾ lb whiting fillet
3 eggs
1½ oz butter (creamed)
mushroom sauce

For panade
¼ pint water
¼ pint milk
2 oz butter
salt and pepper
5 oz plain flour (sifted twice with ½ teaspoon salt)

'Brochet' is French for pike, a fish common to rivers in France. As pike is not readily available here, we suggest turbot, or halibut, as alternatives.

Method

Skin and mince the fish, then weigh. The turbot should weigh 10 oz; the whiting 7 oz.

To prepare the panade: bring the water, milk and butter to the boil with the seasoning. Have the flour ready. When the liquid comes to the boil draw off the heat and at once shoot in all the flour. Beat until smooth. Turn out and allow to get quite cold.

Watchpoint Weigh the panade and the combined fish; they should be equal in weight.

Pound the fish thoroughly, then gradually add the panade, working it well. Add the eggs, one at a time, and the butter. Season and chill overnight.

Watchpoint Once the eggs have been added the mixture will be quite soft, but with the chilling overnight it will be stiff enough to handle.

Divide the mixture into 6–8 even portions, roll lightly to a thick sausage, then carefully drop into boiling salted water and poach for 18–20 minutes. Then lift out with a draining spoon, touching the bowl of the spoon on a piece of absorbent paper to remove any moisture. Cool quickly. Spread the quenelles out on a baking tray to freeze, and when solid, pack them in a rigid plastic container. Make up the sauce and freeze separately.

To use, unwrap the frozen quenelles and place in the oven, pre-set at 350°F or Mark 4, for 30 minutes. Heat the sauce gently, stirring, until thawed; then bring to boiling point. Put the quenelles in a serving dish and pour the hot sauce over them.

Lamb goulash

3 lamb chump ends
1 rounded tablespoon dripping
1 large onion (preferably Spanish)
—finely sliced
1 dessertspoon paprika pepper
1 dessertspoon plain flour
1 small clove of garlic (crushed)
—optional
1 dessertspoon tomato purée
½ pint stock, or water
1 bayleaf
1 large tomato (skinned, sliced and seeds removed)
salt and pepper
2 tablespoons yoghourt
1 tablespoon chopped parsley

Chump ends are cut from the tail end of the loin and, because they have a fairly large proportion of bone, are sold quite cheaply. Braise rather than grill them, and use for making goulash or stews (cooked in the oven or on top of the stove). They can be cooked in the same way as oxtail.

Method

If using the oven, set it at 350°F or Mark 4. Wipe the meat. Heat a stewpan or flameproof casserole, put in dripping and when hot add the meat. Brown for about 4–5 minutes, then add the onion and continue to fry gently for a further 5 minutes. Add paprika, mix well. Cook for 2–3 seconds then draw aside and stir in the flour and add garlic, tomato purée, stock or water, and bayleaf.

Bring slowly to the boil, cover and cook on top of the stove, or in the pre-set moderate oven, for about 50 minutes until tender, and then add the tomato. Adjust seasoning and take out bayleaf. Cool rapidly and turn into a rigid plastic container. Freeze.

To use, turn out into a saucepan and heat gently on top of the stove. Bring to the boil and simmer for 30 minutes, stirring from time to time. Turn into a hot serving dish, spoon over the yoghourt and dust thickly with chopped parsley. Serve with plain boiled potatoes or noodles.

Steak and kidney pie

1½ lb skirt, or sticking, of beef
6 oz ox kidney
salt and pepper
1 tablespoon plain flour
1 shallot, or ½ small onion (finely chopped)
1 teaspoon chopped parsley —optional
½ pint cold water, or stock
hot water, or stock (to dilute gravy)

For flaky pastry
8 oz plain flour
3 oz lard
3 oz butter
about 8 tablespoons cold water
1 egg (beaten)—to glaze

10-inch pie dish, pie funnel

Method

Set the oven at 325°F or Mark 7.

Cut the steak into 1-inch cubes; skin and core kidney and cut into pieces; roll both in well seasoned flour (for this amount of meat add as much salt as you can hold between two fingers and your thumb, and half as much pepper, to the 1 tablespoon of flour).

Place meat in a casserole, sprinkling each layer with the shallot and parsley. Pour in the cold water or stock and cover. Cook in the pre-set oven for 1½ hours. Cool.

Meanwhile prepare the pastry (see page 121). Turn the cold steak and kidney into a foil-lined pie dish and place a pie funnel in the centre. Roll out the pastry ¼ inch thick and cut off a piece large enough to cover and overlap the top of the pie dish; roll the remainder a little thinner and cut two strips, each ¾ inch wide. Damp the edge of the pie dish, press on the strips of pastry and brush with water. Lift the sheet of pastry on your rolling pin and cover the prepared pie.

Watchpoint Do not stretch the pastry when covering pie or it will shrink during cooking and slide into the pie dish. When trimming the pastry to fit the dish, lift the pie on one hand and, holding a knife at an angle away from the dish, cut the overlapping pastry in short, brisk strokes. To trim in one continuous cut would drag the pastry, spoil the appearance and prevent it rising in good flakes.

Seal the edges of double thickness of pastry. Scallop the edge. Any remaining strips or trimmings of pastry can be used to cut a centre decoration of a rose or thistle, and leaves. Wrap the foil over the pastry and freeze. Remove from the pie dish, over-wrap with polythene and seal.

To use, remove the wrappings and replace the pie in the dish. Thaw in the refrigerator for 12–16 hours. Brush with egg beaten with a little salt and bake in the oven, pre-set at 425°F or Mark 7, for about 30 minutes, until the pastry is well browned.

Steak and kidney pie is a traditional English dish, ideal for freezing

Carbonade of beef

1½–2 lb chuck steak
1–2 tablespoons dripping
2 onions (sliced)
1 tablespoon plain flour
1 clove of garlic
½ teaspoon salt
½ pint hot water
½ pint brown ale
bouquet garni
pepper
pinch of grated nutmeg
pinch of sugar
1 teaspoon wine vinegar

> **Carbonade** used to mean a dish that was grilled or boiled over the coals (carbone). It now denotes a rich stew or ragoût (usually of beef) made with beer, and is a characteristic Flemish dish.

Method
Cut the meat into 2-inch squares. Heat the dripping in a flameproof casserole until it's smoking. Put in enough meat to cover the bottom and brown quickly on both sides. Remove these first pieces from the pan if there is any remaining meat to be browned.

Watchpoint If you put too much meat into the casserole, the heat of the dripping is reduced and the meat will stew instead of fry because the juices run. If this happens, the colour of the finished dish will not be rich and brown. Do not overbrown, either, or the meat will become hard.

Remove meat, lower the heat, add the onions and cook until brown. There should be just enough fat to absorb the flour, so pour off a little of the fat if there is too much. Sprinkle in the flour, add the garlic, crushed with ½ teaspoon salt, and return the meat to the pan. Pour the water and brown ale on to the beef, add the herbs, pepper, nutmeg, sugar and vinegar. Stir the pan well to clear any brown juices from the meat that might be stuck on the bottom or sides, and then cover the casserole tightly. Cook gently in the oven for 1½–2 hours at 325°F or Mark 3. Remove the bouquet garni and cool. Turn into a rigid polythene container and freeze.

To use, turn out into a saucepan and heat gently on top of the stove. Bring to the boil and simmer for 30–45 minutes, stirring from time to time. Turn into a hot casserole for serving. Serve with creamed potatoes.

Beef 'en daube'

3–4 lb piece of aitchbone, or topside of beef
1 pigs trotter
pepper (ground from mill)
½ pint brown stock, or water
6–8 oz salted belly pork
¾ lb ripe tomatoes (skinned, seeds removed, flesh chopped)
8 green olives (stoned and sliced)

For marinade
2–3 tablespoons olive oil
1 dessertspoon wine vinegar
½ bottle red wine
1 onion (sliced)
1 carrot (sliced)
1 large bouquet garni (including bayleaf, thyme, parsley stalks and a strip of orange rind)
6 peppercorns
1 clove
½ dozen coriander seeds, or ½ teaspoon ground coriander

This is a dish where the meat is first marinated and then cooked very slowly until tender. There are several variations of a daube but the essential is the long, slow, even cooking. Remember that a daube should be so tender that a spoon – not a knife – is used to cut the meat.

Method

Put all the ingredients for the marinade into a pan, bring slowly to the boil, then draw aside and allow to get quite cold. Place beef in a deep dish and pour over the marinade. Leave for 2–3 days (in warm weather keep meat covered in the refrigerator) turning it over several times.

Then take out the meat and strain the marinade, reserving the vegetables, herbs and spices. Skim the oil from the surface of the marinade and put this into a thick flameproof casserole large enough to hold the beef comfortably; heat, and when the oil is smoking, brown the meat and pigs trotter all over. Draw aside, add the marinade, and the reserved bouquet garni, vegetables, and the spices tied in a piece of muslin. Season with pepper only, and add the stock. Bring slowly to boil, cover and put into the oven, set at 275°F or Mark 1. Leave about 7–8 hours, when the daube should be very tender.

Meanwhile simmer the pork in water for 30–40 minutes, then take up and cut into lardons. Add these to the daube after the first 2 hours' cooking. Add the tomatoes to the daube 1 hour before the end of cooking time.

Take out the trotter, fork the meat off the bone and cut into shreds, return to the casserole with the olives, first taking out the bouquet garni.

Cool and wrap the beef in foil; overwrap with polythene. Put the sauce in a rigid plastic container. Freeze.

To use, loosen the wrappings on the beef and thaw in the refrigerator for 18–24 hours. Thaw the sauce for 2–4 hours. Place the beef in a casserole, pour over the sauce and heat in the oven, preset at 400°F or Mark 6, for about 1 hour.

Ossi buchi

2 lb shin of veal (cut in slices 2 inches thick)
1 oz butter
1 onion (sliced)
1 carrot (sliced)
1 wineglass white wine
$\frac{1}{2}$ lb tomatoes
1 dessertspoon tomato purée
1 clove of garlic (crushed)
$\frac{1}{2}$–$\frac{3}{4}$ pint of jellied bone stock
bouquet garni
1 tablespoon chopped parsley

Method

Brown the veal in the butter and lift carefully out of the pan. Add the onion and carrot, cover the pan and cook over a steady heat, without stirring, for 2–3 minutes. Put the veal back into the pan, making sure that the bones remain upright so the marrow does not fall out as the meat cooks. Pour over the white wine and allow to reduce to half quantity.

Meanwhile scald and skin the tomatoes, cut away the hard core, squeeze to remove a certain amount of the seeds and chop the flesh finely. Add this to the pan with the tomato purée and cook for 10–15 minutes. Then add the garlic, stock and bouquet garni. Cover and cook for $1\frac{1}{2}$ hours.

Take out the veal and set aside to cool. Tip the contents of the pan into a conical strainer and press well.

Watchpoint All the tomato pulp should go through the strainer but the carrots and onion should remain behind.

Reduce this sauce rapidly until syrupy, then cool.

Freeze the veal and sauce separately in rigid plastic containers. To use, thaw in the refrigerator for 8–12 hours, then place the meat in a casserole, spoon over the sauce and heat in the oven, pre-set at 425°F or Mark 7, for about 30–40 minutes. Adjust seasoning and dust with parsley.

Blanquette of veal

2¼ lb breast of veal
2 medium-size carrots (quartered)
2 medium-size onions (quartered)
1 bouquet garni
pinch of salt
1½ pints stock, or water

For sauce
1½ oz butter
3 tablespoons plain flour
1–2 egg yolks
¼ pint creamy milk
squeeze of lemon juice

Traditionally, breast of veal is used for this dish to get a rich, jellied stock from bones. But a greater proportion of shoulder meat can be added, ie. twice as much as breast. Breast of lamb can replace the veal.

Method

Cut meat into chunks (ask your butcher to do this if you are using breast of lamb and also trim off excess fat – otherwise cook in the same way). Soak overnight in cold water, blanch, drain and refresh.

Put the meat into a large pan with the quartered carrots and onions. Add bouquet garni, salt and barely cover with the stock or water. Cover and simmer for 1–1¼ hours until very tender and a bone can be pulled from a piece of meat.

Draw pan aside and pour off all liquid. Cool the veal. The stock should measure 1 pint. If it is more, turn into a pan and boil to reduce to 1 pint.

To prepare the sauce: melt the butter in a separate pan, stir in the flour, cook for 1–2 seconds without letting the butter brown, draw aside and allow to cool slightly. Pour on the stock, blend, then stir until boiling. Boil briskly for 3–4 minutes until sauce is creamy in consistency and then draw aside and cool. Put the meat in a rigid plastic container, pour the sauce into a separate container and freeze.

To use, thaw in the refrigerator for about 12 hours. Turn meat and sauce into a casserole and reheat in the oven, pre-set at 400°F or Mark 6, for about 45 minutes. Mix the egg yolks with the milk in a bowl, add a little of the hot sauce, then pour slowly back into the bulk of the sauce. Adjust seasoning and add the lemon juice. Serve with creamed potatoes or boiled rice.

Braised oxtail

1 oxtail (jointed)
dripping
2 onions (peeled)
2 carrots (peeled and quartered)
3 sticks of celery (cut in 2-inch lengths)
1 tablespoon plain flour
about 1 pint stock, or water
bouquet garni
salt and pepper

Method

Brown the pieces of tail all over in hot dripping in a flameproof dish. Take out; put in the onions, carrots and celery. Leave to brown lightly, then dust in the flour. Remove from heat, add the liquid, bring to the boil, add bouquet garni and seasoning.

Put in the tail, cover dish tightly and cook in the oven at 350°F or Mark 4 for about $1\frac{1}{2}$–2 hours or until tender (when the meat will come easily off the bone). Remove the bouquet garni.

Watchpoint Sometimes oxtail is inclined to be fatty so it is a good plan to cook it the day before, leave to go cold, then skim off the solidified fat before freezing.

Turn into a rigid plastic container and freeze. To use, turn into a casserole while still frozen, and place in the oven, pre-set at 400°F or Mark 6, for about 1 hour, then lower the temperature to 350°F or Mark 4 for about 30 minutes. Adjust seasoning before serving.

Braised oxtail is an economical and simple dish to make – ideal for mothers with large families

Chilli con carne

6 oz red beans (soaked and pre-cooked as directed on the packet)
1 lb minced steak
2 tablespoons oil, or dripping
2 onions (finely chopped)
2 tablespoons chilli con carne spice, or 1 dessertspoon chilli powder
1 dessertspoon paprika pepper

Method

Choose a large stew pan or deep frying pan, heat the dripping in this, add onion and when it is about to turn colour, add the spices. Add the mince, stirring for 4–5 minutes, then add the drained beans and a little of their cooking liquor.

Cover and simmer until beef and beans are tender (about $1\frac{1}{2}$ hours). During this time the pan should be covered and if the mixture gets too thick add a little of the water. The consistency should be that of a rich stew.

Cool and turn into a rigid plastic container. Freeze.

To use, turn into a saucepan and heat gently through on top of the stove for about 1 hour, stirring from time to time.

Casserole of liver

1–$1\frac{1}{2}$ lb lambs liver
6 rashers streaky bacon
$\frac{1}{2}$–1 oz butter, or dripping
4 oz mushrooms
3 medium-size onions (sliced)
1 tablespoon plain flour
$\frac{3}{4}$ pint stock
salt and pepper
12 black olives (stoned)
squeeze of lemon juice

Method

Slice the liver and cut bacon into small pieces. Fry bacon first in butter or dripping in a pan. Take out and put in liver; sauté for 2 minutes on each side until brown. Take out and arrange in a casserole.

Wipe out pan, sauté mushrooms for 3 minutes in a little extra dripping; take out, set aside. Fry onions until golden-brown, add with bacon to liver.

Dust flour into pan, make a roux. Pour on stock and bring to boil. Season, strain into casserole, cover and cook gently for 40–50 minutes in oven at 350°F or Mark 4. Then add mushrooms and olives, finish with lemon juice.

Cool and turn into a rigid container. Freeze.

To use, turn into a casserole and place in the oven, pre-set at 400°F or Mark 6, for about 1 hour. Reduce the temperature to 350°F or Mark 4 and heat for a further 30 minutes. Adjust seasoning before serving.

Puchero bean stew

½ lb red, or brown, beans (soaked overnight in plenty of water)
1 lb salt belly of pork (or beef flank, or brisket)
2 tablespoons salad oil
1 large onion (peeled and sliced)
1 large carrot (peeled and sliced)
1-2 cloves of garlic (chopped)
1 dessertspoon tomato purée
1–1½ pints stock, or water
pepper
bouquet garni
2 caps of pimiento (sliced)
½ lb ripe tomatoes (peeled and halved)
2 saveloys, or equivalent in smoked sausage (about 6 oz)

Puchero (a stew) is a dish of Spanish origin, and there are several versions but this particular recipe comes from South America.

Method

Drain beans, put into a large pan, well cover with cold water and bring slowly to the boil. Simmer for 1½ hours.

Meanwhile put the meat in a separate pan of cold water, bring to the boil and continue cooking for about 1 hour. Then drain both the beans and meat.

Heat oil in the pan in which the beans were cooked, add onion and carrot slices and garlic. Cook for 4–5 minutes, then add beans, tomato purée and stock. Add a little pepper but no salt. Bring pan to the boil, put in meat, add bouquet garni, cover and simmer until beans and meat are tender (about 1½ hours). After 1 hour add the prepared pimiento and tomatoes.

Blanch sausages by putting in cold water, bringing to boil and draining. Then add to the stew and continue to simmer. The puchero should be thick and rich by this time; if too thick, add a little extra stock.

Take out meat and sausages and slice them. Return slices to the puchero.

Cool and turn into a rigid plastic container. Freeze.

To use, turn into a saucepan and heat gently on the top of the stove. Bring to the boil, then simmer for about 1 hour, stirring from time to time.

Scotch collops, or mince

1¼ lb good quality beef (minced)
2 tablespoons beef dripping
1 medium-size onion (finely chopped)
1¼ oz plain flour
¼ pint stock, or water
salt and pepper
1 tablespoon mushroom ketchup (optional)
2 slices dry toast (for garnish)

Method

Melt dripping in a shallow pan. When hot add the onion and, after 2–3 minutes, the minced beef. Stir well with a metal spoon over brisk heat to break up the meat (mince is apt to remain in a solid cake.) Stir in the flour and add the liquid. Season, bring to the boil and cover. Simmer gently for 1½–2 hours, until meat is very tender. Stir occasionally and add a little liquid if necessary. When finished the consistency should be of very thick cream and the colour a rich brown.

Mushroom ketchup can be added if wished.

Cool and turn into a rigid plastic container. Freeze.

To use, turn out the mince into a saucepan and heat gently to boiling point, stirring from time to time. Simmer for 20–25 minutes to heat thoroughly. When ready to serve, make the toast. Remove crusts and cut each round into 8 triangular pieces or sippets. Dish up the mince and surround with the sippets or scatter them over the top. Boiled rice makes a good accompaniment.

Cornish pasties

For dripping crust pastry
8 oz plain flour
pinch of salt
3 oz good beef dripping
about 2½ fl oz cold water

For filling
8 oz good beefsteak
1 large onion (chopped)
2 medium-size potatoes (finely diced)
1 carrot (finely diced)
¼ small swede, or turnip (finely diced)
salt and pepper

The true Cornish pasty should be made with a good-quality raw steak so that all the flavour and juices of the meat are kept inside the pastry. If properly cooked the meat should not be dry. The vegetables may vary; for example, some people do not include carrot. The shape is traditional. A pasty should be slightly curved with two blunt horns at each end, rather like a half moon. This quantity makes 4–6 small pasties.

Method

Sift the flour with salt, rub in the dripping and mix to a firm dough with the water; set aside. Set oven at 400°F or Mark 6.

Cut the meat into very small pieces, and mix with prepared vegetables; season well. Roll out the pastry just over ¼ inch thick and cut out into approximately 5-inch squares. Brush round edges with water and put a good tablespoon of the filling mixture on the top edge of each square. Fold over the plain side of the pastry and press the edges together, curving the pastry downwards to form a crescent shape. Crimp edges slightly to seal well.

Bake in pre-set hot oven for 20 minutes. Then cover with wet greaseproof paper, lower oven to 350°F or Mark 4 and continue to bake for a further 25–30 minutes, when the meat should be well cooked.

Cool and wrap carefully — cooked pastry is very fragile. Either wrap individually in foil and polythene, or place in a single layer in a rigid plastic container — the latter gives the most protection. Freeze.

To use, thaw at room temperature for about 2 hours, then heat through in the oven, pre-set to 400°F or Mark 6, for about 30 minutes (cover the pastry with foil if it starts to over-brown).

Koftas (Meat balls)

1 lb finely minced lamb, or chicken (raw)
½ green pepper
1 medium-size onion
1 clove of garlic (crushed with salt)
¼ carton (2¼ fl oz) plain yoghourt
2 tablespoons chopped coriander
salt
2 teaspoons garam masala
¼ teaspoon ground mace
1 egg (beaten)

Method

Chop the green pepper and onion very finely. Mix all the ingredients with enough beaten egg to bind firmly. Form into balls about 1½ inches in diameter and freeze individually. When solid, pack in twos or fours in foil and overwrap with polythene.

To use, loosen the wrappings and thaw the koftas in the refrigerator for about 12 hours. Turn the frozen curry sauce into a saucepan and heat gently to thaw, stirring from time to time. When thawed, add the koftas and bring to the boil. Simmer for about 40 minutes.

Curry sauce

1 medium-size onion (finely sliced)
2 tablespoons melted butter
2 medium-size tomatoes (skinned and chopped)
2 cloves of garlic (crushed)
2 teaspoons curry powder
1-inch piece of green ginger (scraped and finely chopped)
pinch of cayenne pepper
1 teaspoon salt
2 cartons (½ pint) plain yoghourt
¼ pint hot water (optional)

Method

Fry the onions in the butter until a pale gold. Add the tomatoes, garlic and curry powder. Fry gently for 2–3 minutes. Stir in the ginger, cayenne and salt, then mix the yoghourt and simmer for 15 minutes. If a thinner sauce is required, add the hot water. Cool, turn into a rigid plastic container and freeze.

Garam masala

¾ oz cinnamon
¼ oz cloves
¾ oz brown cardamom seeds
¼ oz black cumin seeds
good pinch of mace
good pinch of nutmeg

Grind ingredients together, or pound them in a mortar, then pass through a fine sieve. Store in an airtight container for up to two weeks.

Beef sauté chasseur

2 lb porterhouse steak, or skirt (cut in one piece)
3 shallots, or 1 small onion
1 clove of garlic
2 tablespoons oil
1½ oz butter
1 tablespoon plain flour
½–¾ pint jellied brown stock
salt and pepper
¼ lb button mushrooms
2 wineglasses white wine
1 dessertspoon tomato purée
1 tablespoon chopped parsley (to garnish)

Method

Cut the meat into 2-inch squares. Chop the shallots (or onion) and garlic very finely. Heat a sauté pan or shallow stew-pan, add the oil and, when hot, drop in 1 oz butter; fry the meat a few pieces at a time until nicely brown on both sides, then remove it from the pan. Reduce the heat, add the shallot and garlic and cook slowly until soft, dust in the flour and continue cooking to a rich russet brown. Draw pan aside, blend in the stock then replace on heat and stir until boiling. Return the meat to the pan, season, cover the pan and simmer until tender (about 45 minutes).

Wash and trim the mushrooms, quarter them or leave whole, depending on their size. Heat a pan, drop in ½ oz butter and mushrooms and cook quickly for about 2 minutes, tip on the wine and boil hard until reduced by half; stir in the tomato purée. Add the mushroom mixture to the meat and continue cooking for about 10 minutes.

Remove the meat from the sauce with a draining spoon. Cool and place them in separate containers and freeze.

To use, thaw the beef in the refrigerator for about 12 hours, the sauce for 2–4 hours. Turn into a saucepan and heat gently, stirring; bring to boiling point and simmer for 30–40 minutes, until thoroughly heated.

Turn into a hot serving dish and dust with chopped parsley. Serve with sauté potatoes and leaf spinach.

Coq au vin

3½–4 lb roasting chicken
4 oz gammon rashers
4 oz button onions
2 oz butter
¼ bottle of Burgundy (7 fl oz)
2 cloves of garlic (crushed with
 ½ teaspoon salt)
bouquet garni
¼–½ pint chicken stock
salt and pepper
kneaded butter

For garnish
1 French roll (for croûtes)—sliced
butter, or oil (for frying)
chopped parsley

Method

First truss the chicken or tie neatly. This is important even though the bird is jointed immediately after browning as it stays compact, making it easy to turn in casserole during browning.

Remove rind and rust from the bacon, cut into lardons (¼-inch thick strips, 1½ inches long). Blanch these and onions by putting into a pan of cold water, bringing to the boil and draining well.

Brown chicken slowly in butter, then remove from casserole. Add onions and lardons; while these are browning, joint chicken and replace joints in casserole. Heat the wine in a separate pan, set it alight and pour it over the chicken while still flaming. (If the wine does not flame, reduce it and then add to chicken.) Add the crushed garlic, bouquet garni, stock and seasoning. Cover casserole and cook slowly for about 1 hour, either on top of stove, or in pre-set oven at 325°F or Mark 3. Test to see if chicken is tender by piercing flesh of the thigh with a fine-pointed cooking knife. If clear liquid runs out, it is ready, if pink continue cooking. When ready remove bouquet garni. Cool quickly, then turn the chicken into a casserole lined with foil. Cover and freeze; remove the casserole and over-wrap the package with polythene.

To use, remove the wrappings and place the frozen chicken in the casserole. Cover loosely and thaw in the refrigerator for 8 hours. Cover and heat in the oven, pre-set at 400°F or Mark 6, for 45 minutes. Thicken slightly with kneaded butter and adjust seasoning. Surround with croûtes and sprinkle with chopped parsley.

If good wine has been used for this dish it needs no accompanying vegetable except creamed potatoes: anything else detracts from the flavour of the sauce

Chicken suprêmes en fritot

5–6 chicken suprêmes
4 oz butter
1 tablespoon seasoned flour
beaten egg
dried white crumbs
juice of ½ lemon
seasoning

For salpicon (to be freshly made)
½ oz butter
12 oz packet frozen sweetcorn kernels (cooked and drained)
pepper (ground from mill)
4 oz ham (shredded)
2 sweet pickled cucumbers (sliced)

1 tablespoon double cream
1 teaspoon chives

If you buy whole chickens use the leg joints for cuisses de poulet Xérès (see page 82).

Cooked chicken pieces ready for freezing, together with other ingredients for chicken suprêmes en fritot

Method

Clarify 3 oz of the butter ready for frying.

Bat out suprêmes to flatten them, brush well with any skimmings from the butter and then roll in the seasoned flour. Brush with beaten egg and roll in the crumbs.

Fry the chicken rather slowly for about 15 minutes in the clarified butter until golden brown on all sides. (If coloured before this, lift the pieces on to a baking tin or rack and place in the oven, set at 350°F or Mark 4, to finish cooking.)

When the chicken is absolutely cold, pack in a foil-lined box or container with as little air space between the joints as possible. Cover and tie in a polythene bag and freeze.

To serve, allow to thaw unopened in the refrigerator for 8 hours; then separate the joints on to a wire rack or baking tin and heat in the oven, pre-set at 350°F or Mark 4, for 20–30 minutes.

Meanwhile prepare the fresh salpicon to serve separately. Melt the ½ oz butter, toss in the sweetcorn, adding ground black pepper to taste. When really hot add ham and cucumber. Bind the sweetcorn with the cream, add chives and dish up.

Place the chicken in a hot serving dish. Heat the remaining 1 oz butter until nut-brown, add the lemon juice and season, and pour it over the chicken while foaming.

Spiced chicken

3 lb roasting chicken
salt and pepper
1 teaspoon ground cumin
½ teaspoon ground allspice
2 teaspoons plain flour
½–1 lb small onions (peeled)
1–2 oz butter
1 dessertspoon tomato purée
½ pint stock

Method

Joint the chicken, season with salt and pepper, rub with the spices and then dust with the flour. Blanch the onions and drain well. Melt the butter in a sauté pan, brown the onions slowly and remove from the pan. Place the joints of chicken in the pan skin-side down and cook slowly until golden-brown; turn them, add the tomato purée and stock and bring to the boil. Replace the onions in the pan, cover and simmer for about 30–40 minutes until the chicken is tender. Place the chicken and onion in a rigid plastic container, and put the sauce in a separate container before freezing.

To use, thaw in the refrigerator for 8 hours. Combine chicken and sauce and heat gently on top of the stove until boiling. Simmer for 20–30 minutes to heat thoroughly.

Cuisses de poulet Xérès

5–6 whole leg joints of chicken
1 shallot
2 oz butter
2–3 tablespoons fresh breadcrumbs
2–3 tablespoons chicken stock
salt and pepper
pinch of ground mace
1 large onion (sliced)
2 large carrots (sliced)
1 stick of celery (sliced)
bouquet garni
$\frac{3}{4}$–1 pint jellied chicken stock

For sauce (to be freshly made)
$1\frac{1}{2}$ oz butter
$1\frac{1}{4}$ oz plain flour
$\frac{3}{4}$ pint stock (see method)
3–4 fl oz double cream

For garnish (to be freshly made)
nut of butter
2 medium-size carrots
1 glass sherry, or Madeira

Method

Cut through the leg joints between the drumstick and thigh and bone out the joints. Trim the meat from the chicken carcass and mince the trimmings together with one portion of thigh. Soften the shallot in 1 oz of the butter and when cold add to the minced chicken and breadcrumbs and pound well. Work in the 2–3 tablespoons stock, a little at a time, and season well with the salt, pepper and mace. Fill this mixture into the boned-out joints and tie with thread.

Melt the remaining 1 oz butter in a flameproof casserole, add the sliced vegetables, cover and cook slowly on top of stove until they begin to colour. Place prepared chicken on top, tuck in herbs and pour over jellied stock. Season well, cover with a buttered paper and bring to the boil. Cover tightly and cook in the oven set at 350°F or Mark 4 for 50–60 minutes.

After braising, remove chicken from the pan, strain the stock, pour this over the chicken and leave to set. Then freeze chicken pieces in the jelly.

To serve, thaw for 8 hours in the refrigerator, turn into a pan and heat on top of the stove for 30–40 minutes. Make the sauce and finish with the garnish.

To make the sauce: melt the butter, stir in the flour and cook slowly until straw coloured. Set aside. Take up the chicken, strain off the stock and reduce to $\frac{3}{4}$ pint. Blend this into the roux and stir until boiling. Leave to simmer until syrupy.

Cut the red part of the carrot in fine strips and put into a small pan with a nut of butter and the sherry. Cover and simmer until the carrot is tender.

Add the cream and carrots to the sauce. Remove the thread from the chicken joints, place them in a hot serving dish and spoon over the sauce. Serve with glazed carrots and plainly boiled rice.

1 Tying up boned and stuffed chicken joints before cooking
2 Cooked, frozen and thawed chicken joints ready to be reheated while the cream sauce and garnish are prepared for the cuisses de poulet Xérès

Jugged hare (Civet de lièvre)

1 hare (jointed)
1 tablespoon dripping
2 onions (diced)
2 carrots (diced)
1 stick of celery (diced)
bouquet garni
1½ pints stock, or water
1 tablespoon redcurrant jelly
1 small glass port wine
1 teaspoon arrowroot, slaked with a little water, or kneaded butter

For forcemeat balls
1 oz butter
1 shallot, or small onion (finely chopped)
1 teacup fresh white breadcrumbs
1 dessertspoon dried herbs
1 dessertspoon chopped parsely
salt and pepper
beaten egg, or milk (to bind)

For frying
seasoned flour
1 egg (beaten)
dried white crumbs
deep fat bath

Method

Marinate hare overnight in a mixture of 4 tablespoons olive oil, 3 tablespoons brandy, 1 sliced onion, 3 parsley sprigs, ½ bayleaf and a sprig of thyme. Then drain. Heat a flameproof casserole and add the dripping. Put in the hare and brown well all over. Take out and put the diced vegetables in the pan. Cover and cook gently for 5–7 minutes. Replace the hare joints on top of the vegetables; add the bouquet garni and pour on the stock. Cover and cook for 1–1½ hours in the oven, pre-set at 325°F or Mark 3. Baste and turn the joints occasionally. Remove the hare and set aside to cool.

Strain the gravy into a pan, skim off the fat and add the redcurrant and port. Boil the gravy and reduce well. Set aside to cool.

Now make the forcemeat balls: melt the butter in a pan, add the onion, cover and cook until soft but not coloured. Mix breadcrumbs, herbs and seasonings together in a basin, add the onion and enough beaten egg or milk to bind. Shape mixture into small balls, roll in seasoned flour, then coat with egg and breadcrumbs. Heat the fat bath to 375°F, lower in the balls with a draining spoon and fry until golden brown. Drain on absorbent paper and cool.

Pour the gravy into a rigid container and pack the hare joints and forcemeat balls separately in foil, over-wrapped with polythene. Freeze.

To use, thaw the hare and forcemeat balls in the refrigerator for 8 hours. Thaw the sauce for 2 hours. Place the hare in a pan, pour over the sauce and heat gently to boiling point. Simmer for 30 minutes. Meanwhile place the balls on a wire rack in a baking tin and heat in the oven, pre-set at 350°F or Mark 4, for 20–30 minutes.

Thicken the sauce slightly with slaked arrowroot or kneaded butter and adjust seasoning. Dish up in a hot casserole, with the forcemeat balls on top.

When jugging hare, it is first marinated and then braised; redcurrant jelly and port are added to the gravy before serving it with forcemeat balls

Casseroled rabbit with mustard

1–2 rabbits
dash of vinegar
salt
$\frac{1}{2}$–$\frac{3}{4}$ lb streaky bacon, or pickled pork (in the piece)
2 tablespoons bacon fat, or dripping
4 medium-size onions
1 tablespoon plain flour
1 pint stock
pepper (ground from mill)
bouquet garni
1 small carton ($2\frac{1}{2}$ fl oz) double cream, or evaporated milk
1 dessertspoon French mustard
1 dessertspoon chopped parsley

Wild rabbit is intended for this dish but, if you prefer, Ostend (tame) rabbit or a cut of shoulder of veal can be substituted.

Method

Trim the pieces of rabbit into neat joints, cutting the wings (forelegs) in two and trimming off the rib-cage. Soak the joints overnight in plenty of salted water with a dash of vinegar to remove the strong rabbit flavour. Then drain joints, rinse and dry thoroughly.

Cut away rind and rust (brown rim on underside) from bacon, or the skin from the pork. Cut into large dice and blanch by putting into cold water, bringing to boil and simmering for 15–20 minutes. Then drain.

Heat dripping in a thick casserole and lightly brown rabbit joints in it. Take out rabbit and put in bacon and quartered onions; fry well until coloured, draw aside, stir in flour and pour on stock. Bring to boil, grind in a little pepper, add bouquet garni and rabbit joints. Cover casserole and cook for $1\frac{1}{2}$ hours, or until rabbit is really tender, in oven at 350°F or Mark 4.

Draw aside, take out bouquet garni and cool. Turn the rabbit and sauce into a foil-lined casserole, cover and freeze. Remove the casserole and overwrap the foil with polythene.

To use, remove wrappings and place the rabbit in the casserole. Place in the oven, pre-set at 400°F or Mark 6, and cook for 1 hour, then lower the temperature to 350°F or Mark 4 for a further 30 minutes. Mix together the cream and mustard and stir into the sauce; adjust seasoning and sprinkle with chopped parsley. Reheat before serving.

Casserole of partridge

3 French partridges
½–¾ pint stock (made with carcasses, root vegetables and bouquet garni)
1–1½ oz butter
1 onion (sliced)
1 carrot (sliced)
1 rasher of bacon (blanched and diced)
1 dessertspoon plain flour
bouquet garni
¼ lb chipolata sausages
4–6 croûtes of fried bread

For stuffing
1 shallot (finely chopped)
1 oz butter
3 oz fresh white breadcrumbs
1½ oz raisins (stoned)
1 oz chopped walnuts
1 teaspoon finely chopped parsley
1 small egg (beaten)
salt and pepper

Trussing needle and fine string, or poultry pins

Method

Ask the poulterer to bone out the partridges, leaving the leg bones in. Spread out birds on your work surface. Clean the carcasses, break them up and use to make the stock, with root vegetables and bouquet garni to flavour.

To prepare the stuffing: soften shallot in the butter, then mix with the remaining stuffing ingredients, binding mixture with the beaten egg and seasoning to taste. Spread stuffing on the partridges and sew up with fine string or secure with pins.

Heat butter in a flameproof casserole, and brown the partridges; add the onion, carrot and bacon. Cook for 2–3 minutes, then dust with flour, add bouquet garni and stock. Cover casserole tightly and braise slowly for 45–60 minutes in the oven, pre-set at 325°F or Mark 3.

Remove birds from casserole. Split them in half and cool. Wrap in foil with a piece of freezer paper between each portion and over-wrap with polythene. Pour the sauce into a rigid container. Freeze.

To use, thaw the partridge for 24 hours in the refrigerator. Thaw the sauce for 2–4 hours. Set the oven at 400°F or Mark 6. Place the partridge in a casserole, pour over the sauce and put in oven for 30 minutes. Heat a little butter in a roasting tin, put in the sausages, place in the oven to brown. When the sausages are brown, cut them in half diagonally and add to the casserole.

Serve each half of partridge on a croûte of fried bread, with the sauce and sausages.

Chicken and tongue rolls

4–6 oz cooked chicken
4–6 oz cooked tongue
2–3 tablespoons mayonnaise
6–8 slices cooked ham
little beaten egg, or water

For puff pastry
6 oz butter
6 oz plain flour
pinch of salt
cold, or iced, water

The following recipe is good for children's picnics. It will make 6–8 rolls according to the number of slices of ham.

Method
Prepare the pastry and chill.
Shred the chicken and tongue finely, bind with the mayonnaise. Spread this mixture on the slices of ham and roll up like a fat cigar. Roll out the pastry about $\frac{1}{4}$ inch thick; cut into rectangles and place a roll of ham on each; fold over pastry and seal (leaving the ends open) with a little beaten egg or water.
Roll out the pastry trimmings and cut into narrow strips. Brush the rolls lightly with water and lay the strips over them to decorate. Wrap in foil and then put into a polythene bag for freezing.
To serve thaw out overnight in the refrigerator and then bake in the oven, set at 425°F or Mark 7, for 15–20 minutes.

Madras chicken curry

6–8 chicken drumsticks and thighs
1 oz butter
1 onion (finely chopped)
1 tablespoon coriander
1 teaspoon each cumin and chilli powder
$\frac{1}{2}$ teaspoon turmeric
1 dessertspoon tamarind (infused in $\frac{3}{4}$ cup water)
1 dessertspoon coconut cream

Method
Brown the chicken joints slowly in the butter. Take out, add the onion and, after 2–3 minutes, the dry spices. Fry for a few seconds, then add the strained tamarind liquid and the coconut cream. Stir well and replace the chicken. Cover tightly and simmer for 35–40 minutes. Remove the chicken and reduce the gravy if necessary (this curry should be dry).
Cool and turn into a foil-lined casserole. Cover and freeze. Remove the casserole and overwrap the package with polythene.
To use, remove the wrappings and place the curry in the casserole. Thaw in the refrigerator for 8 hours. Place in the oven, pre-set at 400°F or Mark 6, for about 45 minutes.
Serve when reheated with the usual accompaniments of freshly boiled rice and fresh chutneys.

Terrine of chicken

3½ lb roasting chicken
salt and pepper
pinch of ground mace
juice of ¼ lemon
½ lb cooked ox tongue (see method, page 93)
1 cup (3 oz) fresh breadcrumbs
1 small onion (finely chopped)
1 teaspoon fresh chopped sage
1 egg yolk
3–4 tablespoons stock
6 oz ham (finely sliced)

6-inch diameter top (No. 2 size) soufflé dish, or cake tin

Method

Remove the skin from the chicken, cut off the suprêmes and cut in thin slices. Season them with salt, a little ground mace and a squeeze of lemon, and then set on one side while preparing the other ingredients.

Mince the tongue and all the dark meat from the chicken carcass; mix with the crumbs and onion, season with salt and pepper and add the sage. Mix the egg yolk and stock together and work into the minced mixture a little at a time.

Line the soufflé dish (or cake tin) with the sliced ham. Cut any left over into thin shreds and mix with the sliced white chicken meat.

Fill the dish with alternate layers of minced mixture and shredded meat, beginning and ending with the former. Cover the dish with foil and steam for 1 hour. Allow the terrine to cool in the dish, cover with a plate and put a 2 lb weight on the top. When cold turn out of the dish, wrap in foil and polythene and freeze.

To serve, allow to thaw in the refrigerator overnight.

Terrine of duck

1 large duck (weighing 5–6 lb)
pepper (ground from mill)
1 glass port, or golden sherry
½ teaspoon allspice
1 lb pork, or veal (minced)
½ lb lambs, or pigs, liver (minced)
1 medium-size onion (finely chopped)
1 bayleaf
pared rind of ½ an orange
little melted lard

Oval terrine (2 pints capacity)

Method

Skin the duck, slice off the breast meat and cut into shreds; grind over a little pepper, add the port (or sherry) and allspice. Leave to marinate.

Set the oven at 325°F or Mark 3. Cut away all the meat from the legs and carcass of the duck. Mix with the minced pork (or veal), liver and onion. Work together until smooth, then layer this farce in a terrine with the breast meat of the duck. Smooth the surface, press a bayleaf and the orange rind on the top. Cover the terrine, and spread a luting paste round the edge (see page 46).

Cook in a bain marie in the pre-set oven for 1–1½ hours. Then remove lid, press and leave until cold. Run a little melted lard over the top. Turn out of the dish, wrap in foil and a polythene bag and freeze.

To serve, allow to thaw at refrigerator temperature overnight. Serve with orange and brazil nut salad.

Spiced beef 1

5 lb silverside of beef

For spice mixture
2 cloves of garlic
3 oz soft brown sugar
1 oz saltpetre
1 oz allspice (crushed or pounded)
2 bayleaves (chopped or powdered)
4 oz salt

For cooking
2 onions (quartered)
2 carrots (quartered)
1 stick of celery (sliced)
large bouquet garni
cold water

Piece of butter muslin

There are two types of spiced beef. The first one is a piece of beef rubbed well with spices and seasoning and left for 7–10 days before cooking. The other is beef bought salted and cooked with vegetables and herbs to flavour. Both types are pressed and glazed.

Method

Peel garlic and split each clove into 3–4 pieces, make small incisions in the meat with the point of a knife. Mix other spice ingredients together in a bowl and rub well over meat (saltpetre gives meat a pinkish colour). Keep in a deep dish in a cool larder or refrigerator for a week, rub meat well every day with spice mixture.

Then take out meat, tie it in muslin and put into a large pan with the prepared vegetables and bouquet garni. Cover well with cold water, put lid on the pan and simmer for 3–4 hours until meat is tender.

Cool slightly in the liquid, then take out meat, remove muslin, and set meat in a deep dish with an enamel plate on top of it with at least a 4 lb (or equivalent) weight on it. Leave overnight.

Wrap with foil and over-wrap with polythene. Seal and freeze.

To use, thaw in the refrigerator for 18–24 hours.

Spiced beef 2

4–5 lb salted brisket of beef
10 peppercorns
large bouquet garni
2 onions (quartered)
2 carrots (quartered)
1 stick of celery (cut in four)
cold water

Method

Cut beef in two and put in a large pan. Add peppercorns, the bouquet garni, prepared vegetables and water to cover. Simmer until meat is tender (about 3–4 hours), then lift out into a deep dish.

Put one piece of beef on top of the other and run a short thin skewer vertically through each end, pressing them down to level with top of meat.

Put an enamel plate on top of meat with at least a 4 lb (or equivalent) weight on it. Leave overnight.

Freeze as for Spiced beef 1.

Pork brawn

$\frac{1}{2}$ pig's head (salted)
1–1$\frac{1}{2}$ lb shin of beef
1 large onion (peeled)
1 large bouquet garni
cold water
black pepper (ground from mill)

It is difficult to cut down on the quantity given above, so divide into 2 portions before freezing.

Method

Rinse the pig's head in cold water and put into a large pan with the beef, peeled onion (left whole) and bouquet garni. Barely cover with cold water, cover pan closely with a tightly-fitting lid, bring to the boil and then simmer for 2$\frac{1}{2}$–3 hours or until both head and shin are very tender.

The head is cooked when the bones can be pulled out easily. Cool slightly. Lift meat on to a large dish, strain stock and return to pan, boil gently.

Meanwhile remove bones and pull the meats into pieces with two forks. Pepper well, and divide the meat between two rigid plastic containers. Ladle in enough of the reduced stock to come level with the meat.

Chill and freeze. To use, thaw in the refrigerator for 18–24 hours. Turn out, slicing fairly thinly to serve. A sharp dressing or sauce (Cumberland or mustard and brown sugar) is good with most rich meats.

Raised pork pie

1 lb pork (lean and fat mixed)
salt and pepper
1 rounded teaspoon mixed dried herbs
¼ pint jellied stock (made from pork bones)

For hot water crust
1 lb plain flour
1 teaspoon salt
7 oz lard
¾ pint milk and water (mixed in equal proportions)
milk (for glaze)—optional

1 jar (eg. kilner jar)

Method

Dice pork for filling, season well and add herbs.

For hot water crust: warm a mixing bowl and sift in flour and salt, make a well in the centre of the flour.

Heat lard in milk and water. When just boiling, pour into the well in the flour, stir quickly with a wooden spoon until thick, then work with the hand to a dough. Turn on to a board or table, cut off a quarter of the dough, put it back in the warm bowl and cover with a cloth.

Pat out the rest of dough with the fist to a thick round, set a large jar in the centre and work dough up sides.

Watchpoint You must work quickly and mould pastry while it is still warm, otherwise lard sets and pastry becomes brittle.

Let dough cool then gently lift out jar. Fill dough case with meat mixture. Roll or pat out remaining dough to form a lid, leave a small hole in it, then put on top of pie, seal edges. Glaze with milk if wished.

Slide pie on to a baking sheet and bake in pre-set oven for 1–1½ hours at 350°F or Mark 4. If pie is getting too brown, cover with damp greaseproof paper towards end of cooking time. Leave till cool before placing a funnel in hole in lid and filling up with jellied stock.

When cold, wrap in foil, overwrap with polythene and freeze.

To use, thaw in the refrigerator for 18–24 hours, loosening the wrappings first.

To make a raised pork pie case: work dough up sides of jar while it is still warm

Pressed tongue

1 ox tongue (4½–6 lb)
cold salted water
1 large carrot
1 large onion (peeled)
6–8 peppercorns
1–2 bayleaves

Tongue press (available in two sizes), or round cake tin, or soufflé dish

Method

Put the tongue into a large pan, cover well with cold salted water. Bring slowly to the boil, then add other ingredients. Cover pan and simmer gently for 4–5 hours, or until the tongue is very tender.

Test by trying to pull out the small bone at the base of the tongue – if it comes away easily the tongue is done. It is also advisable to stick the point of a knife into the thickest part (just above the root). It will slip in easily if the meat is cooked.

Watchpoint It is almost impossible to overcook a tongue so you should always give it the benefit of the doubt. If even slightly undercooked it can be rubbery.

Cool in the liquid, then lift out the tongue and put in a bowl of cold water. This will make it easier to handle.

Peel off the skin, cut away a little of the root and remove any bones. Curl the tongue round and push into the tongue press, cake tin or soufflé dish. The tongue must fit closely into the tin or dish. Press down well, using a small plate with a weight on top, if no press is available.

Chill overnight, then turn out, wrap in foil and overwrap with polythene. Freeze.

To use, thaw in the refrigerator for 18-24 hours, loosening the wrappings first.

1 *Press the tongue well down into the tin so that it fits snugly*

2 *Make sure that the press on the tongue is as tight as possible*

Terrine of hare

1 hare
4 oz fat bacon (in the piece)
1 wineglass port, or sherry
good pinch of ground allspice
pepper (ground from mill)
2 shallots (finely chopped)
8 oz pork (minced)
8 oz sausage meat
1 dessertspoon chopped mixed herbs
6 oz tongue (in 2 slices)
1 bayleaf
strong jellied stock (made from hare bones with 1–2 veal bones)
luting paste (see page 46)

Method

Reserve the liver and blood of the hare. Lift the fillets from the back of the hare with a sharp knife and cut into strips. Cut bacon into strips. Lay both in a dish, pour over the port or sherry, add spice, pepper and shallots. Cover, leave overnight.

Cut off all the meat from the rest of the hare, mince with its liver and add to the minced pork and sausage meat. Add herbs. Season farce well and pour in any liquid from hare. Cut the tongue into strips and add to the hare fillets.

Press a third of the minced stuffing into a terrine or oven-proof casserole and spread half of the marinated meat and tongue on top. Cover with a half of remaining farce, then rest of the meat and, finally, the remaining farce. Smooth top, press a bayleaf in centre, cover, seal lid with luting paste.

Cook in a bain maire in the oven at 325°F or Mark 3 for 2 hours. Remove lid, press overnight (using about a 4 lb weight), then fill up with stock. When jellied, turn out and wrap in foil. Overwrap with polythene and freeze.

To use, remove the wrappings and return the frozen terrine to the dish in which it was cooked. Thaw in the refrigerator for 18–24 hours and bring to room temperature for serving.

1 *Layer the liver stuffing and shredded game in terrine or casserole; start and end with stuffing*
2 *Smooth the top and add the bayleaf. Then seal lid with a little of the luting paste*

Chip potatoes

1½ lb even-size potatoes
 (weighed when peeled)
deep fat, or at least 1-inch depth of fat in frying pan

Method

After peeling, square off ends and sides of potatoes, cut in ½-inch slices, then into thick fingers. Soak in cold water for 30 minutes, then drain, Wrap in absorbent paper or cloth and leave for 20–30 minutes. Heat fat, dip in basket; when fat reaches 350°F gently lower the basket, full of potatoes, into fat. If you do not have a thermometer, drop in a finger of potato; if this sinks to bottom of pan and fat starts to bubble gently, fat is ready. Fry gently until potatoes are just soft but not coloured. Lift out and drain, still in basket, on absorbent paper. Cool and pack in polythene bags. Freeze.

To use, heat fat to 360–375°F; carefully lower in basket, fry chips to a deep golden-brown. Drain well on absorbent paper, turn into a hot dish for serving and sprinkle with salt. Potatoes double-fried in this way are crisply tender on the outside and evenly browned. When cooking fish and chips, fry potatoes first so that there is no chance of crumb coating from fish spoiling the fat for the potatoes.

Creamed potatoes

1½ lb potatoes
1–2 oz butter
¼ pint milk
salt and pepper

Method

Cut the potatoes in even-size pieces, if very large, and cook in salted water until tender. Do not test with the thick prongs of a table fork or the potato will break.

Watchpoint Don't let water boil away from potatoes; they must cook in the water.

When potatoes are tender, tilt the lid of the pan and pour off the water. Return to a gentle heat and, with the lid half-on, continue cooking for 2–3 minutes until the potatoes are dry. Then add butter – however much you like – season and crush the potatoes with a potato masher or a fork. Pour on ¼ pint of boiling milk (enough for 1½–2 lb potatoes). Beat well. Cool. Pack in rigid containers or pipe or spoon out into small portions on a baking tray. Freeze. Place frozen scoops of potato on a paper or foil plate and wrap with polythene.

To use, place on a baking tray, cover loosely with foil and bake at 400 F or Mark 6 for 30 minutes.

Cauliflower au gratin

1 large cauliflower
1 bayleaf
3 tablespoons grated cheese
3 tablespoons fresh white breadcrumbs
½ oz butter (melted)

For mornay sauce
1½ oz butter
3 tablespoons plain flour
¾ pint milk
salt and pepper
3 tablespoons grated cheese
French, or English, mustard

When cooking cauliflower, add a bayleaf; this lessens the strong smell and gives a delicate flavour to the vegetable.

Method

Wash cauliflower thoroughly in salted water. Trim stalk but leave some green leaves on. Then break into sprigs (if necessary, use a knife to cut the stalk so that it remains attached to the sprigs and is not wasted) and boil for about 15 minutes, or until tender, in salted water with the bayleaf.

Meanwhile, prepare mornay sauce. Melt butter in a pan and stir in flour off the heat. Blend in milk, then stir until boiling. Cook for 2 minutes, season, draw on one side and cool before beating in the cheese by degrees. Then stir in the mustard to taste.

Now carefully drain cauliflower, butter a basin and arrange sprigs in it with the stalks towards the centre. When the basin is full, spoon in 2–3 tablespoons of sauce. Press down very lightly to bind the sprigs together and then invert the basin on to a foil or paper plate. Cool and freeze. Wrap in polythene. Freeze remaining sauce separately.

To use, remove polythene, cover loosely with foil and bake in the oven, pre-set at 400°F or Mark 6, for 30 minutes. Meanwhile turn out the sauce into a small pan and heat gently on top of the stove. Spoon the sauce over the cauliflower. Mix together the grated cheese and breadcrumbs and scatter over the sauce. Sprinkle well with melted butter and return to the oven for 10 minutes to brown.

Sliced peaches

It is best to use Hale peaches for deep-freezing purposes. First skin the peaches and then slice them straight into a small airtight container, layering them with sugar at the same time. Pack the layers very tightly, so that you need a certain amount of pressure to put on the lid.

Leave the covered container upside-down in the refrigerator overnight, or for 12 hours: this will enable the sugar to melt and the juice to run from the peaches. When you put the container into your home-freezer you will find that the peaches have sunk to leave enough headspace in the container.

This method of freezing peaches prevents discolouration which is liable to occur when freezing this kind of fruit in a heavy syrup.

Raspberries in strawberry cream

8–12 oz raspberries
$\frac{1}{4}$ pint of frozen sweetened strawberry purée
$\frac{1}{4}$ pint double cream
crystallised violets (for decoration)

Method

Arrange the raspberries in four individual containers. Whip the cream lightly until it just begins to thicken, then add the strawberry purée, a little at a time, and continue whipping until the cream holds its shape. Spoon this over the raspberries and decorate with crystallised violets. Cap each container with foil and freeze. Overwrap with polythene.

To use, thaw in the refrigerator for 6 hours.

Strawberry purée

To make a sweet strawberry purée suitable for deep-freezing: rub the strawberries through a nylon sieve. Add icing sugar or caster sugar to taste. Put the purée into an airtight container, leaving a little headspace in the container. Put the container straight into the freezer.

Strawberry charlotte

7½ fl oz strawberry purée
3 eggs
2 yolks
6 oz caster sugar
½ pint double cream
½ oz gelatine
juice of ½ lemon (made up to 2½ fl oz with water)

To finish
¼ pint double cream
langues de chats biscuits
angelica (cut into diamonds)

7½–8 inch diameter shallow cake tin (*preferably spring form or loose bottom*)

Method

Put the eggs, yolks and sugar in a bowl and whisk over gentle heat (or without heat, using an electric mixer at high speed) until thick and mousse-like. Half whip the cream and dissolve the gelatine in the lemon juice and water over gentle heat. Add the strawberry purée, gelatine and cream to the mousse and stir over ice until it thickens creamily.

Pour into the cake tin, lined with foil, cover and freeze. The following day carefully turn it out of a tin and wrap in more foil and a polythene bag and return to the freezer.

To serve – turn while frozen on to a serving plate; whip the cream and sweeten very lightly. Spread the biscuits with cream and press overlapping around the charlotte, decorate with rosettes of cream and diamonds of angelica. Allow 4–5 hours in refrigerator or cool larder to thaw before serving.

Tangerine sorbet

juice of 6 tangerines (to give ½ pint)
½ oz gelatine
juice of 1 lemon (strained and made up to 2½ fl oz with water)
7 oz lump, or granulated, sugar
16 fl oz water
1 egg white
caster sugar

Method

Soak the gelatine in the lemon juice and dissolve over heat, allow to cool and then add to the tangerine juice. Dissolve 4 oz sugar in ½ pint water and boil steadily to 220°F, using a sugar thermometer (this should take about 8 minutes). Stop the boiling, then leave to get quite cold. Mix the fruit juice and sugar syrup together, pour into a suitable container, place in the freezer for 2 hours. Beat the egg white until stiff. Take out the tangerine mixture, whisk well and fold in the egg white; cover and return to the freezer for 2–3 hours.

Meanwhile boil the peel of 2 tangerines for 15 minutes in lightly salted water; drain and cut into julienne strips. Prepare a sugar syrup with the remaining 3 oz lump (or granulated) sugar and the rest of the water, and cook the tangerine peel in the syrup gently until almost transparent. Lift on to a rack or sieve to drain, then roll in caster sugar and leave to dry. Serve the sorbet in glasses with the candied peel on the top.

If sorbet is left more than 2–3 hours in the freezer, transfer it into refrigerator 1 hour before serving.

Chestnut parfait

1 can (16 oz) sweetened chestnut purée
½ pint custard (made with 4 egg yolks, 1 oz sugar, ½ pint milk and 1 vanilla pod)
1 pint double cream
3–4 tablespoons maraschino
¼ lb packet candied chocolate orange sticks (cut in pieces)
double cream (whipped) – for masking
marrons glacés debris (optional)

7–8 inch diameter cake tin, or fluted jelly mould (2½ pints capacity)

Method

Make the egg custard, using the vanilla pod for flavouring, and cool; whip the cream until just thickening, then add the maraschino. Blend the chestnut purée, custard and cream together, add about half the orange sticks.

Turn into the mould, cover with foil and freezer paper, and tie in a polythene bag and freeze.

Note: if being deep-frozen for some time before a party, allow to thaw out at refrigerator temperature for 1 hour before serving.

Serve masked with cream and decorated with marrons glacés debris (broken pieces) and the remaining candied orange sticks.

Peach gâteau with mousseline sauce

4–6 frozen peaches (see page 98)

For genoese pastry
3 oz plain flour
pinch of salt
1½ oz butter
3 eggs
3¼ oz caster sugar

For mousseline (to be freshly made)
3 egg yolks
1 tablespoon caster sugar
2 tablespoons Marsala, or golden sherry
3–4 tablespoons of lightly whipped double cream

8-inch diameter sandwich tin

Method

Transfer the peaches from the freezer to your refrigerator 12 hours before you need them. Prepare the genoese pastry (see page 125) and bake in the tin in a pre-set oven at 350–375°F or Mark 4–5 for 30–35 minutes. (This cake can, of course, be made in advance and deep-frozen.)

Meanwhile, make the mousseline. Place the egg yolks, sugar and Marsala (or golden sherry) in a bowl and whisk over hot water until thick. Continue whisking until quite cold, then add the cream.

When the cake is quite cold cover the top with the peaches, then cover with the mousseline.

Peach croissants with iced zabaione

½ pint carton of frozen peaches (see page 98)
6 croissants (see page 105)
1 oz butter

For zabaione (freshly made)
2 oz sugar
1 tablespoon water
1 egg white
3 egg yolks
1 tablespoon Marsala, or golden sherry

Method

Transfer the peaches from the freezer to the refrigerator 12 hours before you need them.

To make the zabaione, put the sugar and water into a small pan and dissolve the sugar slowly over low heat; then boil up quickly to 250–260°F, or until a little of the syrup will form a hard ball when dropped into cold water. Beat the egg white until stiff, add the sugar syrup and mix quickly with a whisk until thick. Place the egg yolks and Marsala (or sherry) into a bowl and whisk over heat until thick and mousse-like; combine with the meringue mixture and chill thoroughly. Melt the butter in a frying pan, add about 2–3 tablespoons of the syrup from the peaches and allow to bubble together for 2–3 minutes. Fry the croissants in this mixture until golden-brown, remove from the pan, arrange around a serving dish, keep warm. Put peaches in the middle of the dish of croissants and pour over the zabaione.

Danish pastries

Danish pastries are made in various traditional shapes — the most usual are cartwheels and pinwheels. They are filled with almond paste, jam, sultanas and raisins, apples, etc.

Basic recipe

12 oz plain flour
large pinch of salt
1 oz yeast
2 oz caster sugar
1 teacup lukewarm milk
9 oz butter
1 egg (beaten)
little extra egg (beaten) — for brushing
glacé icing

This quantity will make 12 pastries of any shape.

Method

Sift the flour with the salt into a mixing bowl, cream the yeast with the sugar until liquid, add a good teacup of lukewarm milk and 2 oz of the butter, stir until dissolved; then add the beaten egg. Pour these liquid ingredients into the flour and mix to a smooth dough. Cover the dough and leave at room temperature for about 1 hour or until double in bulk. Punch down the dough, turn it on to a floured board and knead lightly. Roll out to an oblong and cover two-thirds of the dough with half the remaining butter, divided in small pieces the size of a walnut. Fold and roll as for flaky pastry. Fold in three and roll again. Put on the remaining butter, cut into pieces, fold and leave for 15 minutes. Roll and fold twice more and leave again for 15 minutes. Chill the dough for a little while.

Then roll pastry until it is $\frac{1}{2}$ inch thick, shape as described below, prove and brush with beaten egg and bake in a hot oven, set at 400°F or Mark 6 for 25 minutes.

Cartwheels

Roll out the Danish pastry dough as thinly as possible to a large oblong, spread carefully with a very thin layer of almond filling, then sprinkle with raisins and roll it up as for a swiss roll. Cut the roll into $\frac{1}{4}$-inch slices, and place the slices, cut side down, on a greased baking tin. Prove, brush with beaten egg and sprinkle flaked almonds on the top before baking.

Pinwheels

Roll out dough thinly and cut into 4-inch squares. Cut the dough from each corner to within a $\frac{1}{2}$ inch of the centre. Fold four alternate points to the centre, pressing them down firmly. Put a little jam or almond filling in the centre, then prove and bake.

Crescents

Roll dough into a large circle $\frac{1}{8}$ inch thick and cut it into triangles or wedges. Pour a little almond filling on each triangle and roll them up loosely, starting at the base of the triangle, and then shape into crescents. Prove and bake.

Envelopes

Roll out the dough thinly and cut into 4-inch squares. Spread with vanilla cream and fold the corners in towards the middle. Press the edges down lightly. Prove, then bake for 12–15 minutes in a hot oven set at 400°F or Mark 6.

Combs

Roll out the dough fairly thinly and cut into strips about 5 inches wide. Place an apple or almond filling in the middle and fold both sides over. Brush lightly with beaten egg and roll in crushed lump sugar and chopped almonds. Cut into pieces about 4 inches long and gash about four or five times on one side; open out the slits slightly. Prove, then bake for 12–15 minutes in a hot oven at 400°F or Mark 6. These combs can be brushed slightly with beaten egg before baking to give them a glazed finish.

Vanilla cream

1 tablespoon plain flour
1 teaspoon cornflour
1 egg yolk
1 tablespoon sugar
$\frac{1}{4}$ pint milk
2–3 drops of vanilla essence

Method

Work the flours, egg yolk and sugar together, adding a little milk. Bring the rest of the milk to the boil, pour on to the mixture, blend and return to the pan. Stir until boiling. Allow to cool then flavour with a few drops of vanilla essence.

Almond filling

2 oz almonds (ground)
2 oz caster sugar
little beaten egg

Method

Mix the almonds and sugar together and bind with enough egg to bring to a firm paste.

Apple filling

1 lb cooking apples
$\frac{1}{2}$ oz butter
grated rind and juice of $\frac{1}{2}$ lemon
3–4 tablespoons granulated sugar

This can be used for any shape of Danish pastries and the pastries can either be finished with a soft icing or brushed with a little apricot glaze.

Method

Wipe the apples, quarter and core them, but do not peel. Rub the butter round a saucepan, slice in the apples and add the grated rind and lemon juice. Cover and cook them slowly to a pulp.

Rub pulp through a nylon strainer, return to the rinsed-out pan with the sugar. Cook gently until thick. Turn out and allow to get quite cold before using. This can be turned into a jam jar and used as required.

To freeze:

Pack the pastries in a single layer in foil, polythene bags or rigid containers and freeze.

Danish pastries continued

To use, loosen the wrappings and thaw at room temperature for 1½ hours. Refresh in the oven, pre-set at 350°F or Mark 4, for 5 minutes. Remove from the oven and brush with glacé icing. Cool before serving.

Folding dough into three, with un-buttered portion first

A selection of Danish pastries: top left to right: cartwheels, envelope, comb; centre: pinwheels; bottom: crescents

French croissants

8 oz plain flour
¼ oz yeast
2–3 tablespoons tepid water
½ teaspoon salt
1 tablespoon sugar
4 oz butter
about 3 tablespoons milk
1 egg (beaten)

This quantity will make 18 croissants.

Method

Sift the flour. Dissolve the yeast in the water and mix with a quarter of the flour to make a small ball of dough. Cut a cross on top and drop this yeast cake into a large bowl of warm water.

Meanwhile mix the remaining flour with the salt, sugar, half the butter and sufficient milk to give a soft, but not slack, dough. Beat the dough on the pastry board until it is smooth and elastic. When the yeast cake has risen to the surface of the water and is almost double its size, drain carefully and mix into the dough thoroughly. Put the dough into a floured bowl, cover and leave overnight in a cool larder or refrigerator.

Shape the remaining butter into a flat cake. Turn the dough on to a floured board, roll it out to an oblong and place the butter in the centre. Fold one third of the dough over the butter and fold the other third of the dough on top to make 3 layers. Turn the folded dough so that one of the open ends faces front. Roll out again, fold over as before and turn. Repeat once more. Wrap in a cloth and leave for 15 minutes in a cool place. Repeat the rolling and folding twice more.

To shape the croissants: roll the dough to an oblong $\frac{1}{8}$ inch thick, divide lengthways and cut each strip into triangles. Roll up, starting from the base, curl and place on a lightly floured baking sheet. Cover with a cloth and prove in a warm place for 15–20 minutes. Brush with beaten egg and bake in a hot oven, set at 425°F or Mark 7, for 5 minutes, then reduce the heat to 375°F or Mark 5 and continue cooking for about 10 minutes, until the croissants are browned. Cool, then pack in a single layer in foil, polythene bags or a rigid container. Freeze.

To use, thaw at room temperature for $1\frac{3}{4}$–2 hours, then remove the wrappings, cover loosely with foil and refresh in the oven, pre-set at 425°F or Mark 7, for 5 minutes.

Victoria sandwich

about 6 oz butter
about 6 oz caster sugar
3 large eggs
about 6 oz self-raising flour
pinch of salt
1–2 tablespoons milk

To finish
3 tablespoons warm jam, or lemon curd
caster sugar (for dredging)

Deep 8-inch diameter sandwich tin

To make a good Victoria sandwich, weigh eggs in their shells and use exact equivalent of butter, sugar and flour.

Method
Grease and line sandwich tin; set the oven at 350°F or Mark 4.

Using the creaming method, soften the butter in a bowl, add the sugar and cream them together until soft and light. Whisk the eggs, add a little at a time and then beat thoroughly. Sift the flour with the salt and fold into the mixture a third at a time, adding enough milk to make the mixture drop easily from the spoon. Spread the mixture in the prepared tin and bake in pre-set oven for about 40–45 minutes.

To test if cake is ready press lightly with fingertips and it should spring back. The colour should be golden-brown, and the cake shrunk from sides of the tin. Have two wire cooling racks ready, and put a folded clean tea towel or double thickness of absorbent paper on one of them. Loosen the sides of the cake with a round-bladed knife, place the rack with the towel or paper on top of the cake (towel next to it) and turn over; remove the tin and disc of paper from the base. Place second rack on top of cake base and carefully and quickly turn it over again. This prevents the cake having the marks of cake rack on its top.

Cool, wrap in polythene and freeze. To use, thaw at room temperature for 1–2 hours; split in half, fill with jam or lemon curd; dust top with caster sugar.

English madeleines

Victoria sandwich mixture with about 2 eggs and equivalent weights of butter, caster sugar and self-raising flour (about 4 oz of each)—see left
2–3 drops of vanilla essence
4 tablespoons apricot, or redcurrant glaze
6 tablespoons dessicated coconut
8 glacé cherries

14–16 dariole moulds, or castle tins

Method

Grease moulds or tins well, and dust with flour. Set the oven at 375°F or Mark 5.

Prepare sponge mixture as for Victoria sandwich, flavour with vanilla essence and half fill the moulds or tins. Bake for 8–10 minutes until golden-brown.

When the cakes are cool, trim the tops to give them a flat surface when inverted. Turn them upside down and spear each separately on a fork. Then brush with warm glaze, roll at once in the desiccated coconut, decorate with half a glacé cherry.

Pack in a single layer in a rigid container or cardboard box and freeze.

To use, thaw at room temperature for about 1 hour.

Rock cakes

8 oz self-raising flour
pinch of salt
4 oz butter
3 oz caster sugar
3–4 oz sultanas
1 oz candied peel (finely chopped, or shredded)
2 eggs
1–2 tablespoons milk

Method

Grease a baking tin and set the oven at 425°F or Mark 7.

Sift the flour with the salt into a bowl, add the butter, cut it into the flour with a table or palette knife. Then using your fingertips, rub it in until the mixture resembles fine breadcrumbs. Stir in the sugar, sultanas and peel.

Whisk the eggs to a froth and add them to the dry ingredients with a fork, adding as much milk as necessary to bind the dry ingredients together. Put out in tablespoons on to the prepared tin and bake at once in pre-set oven for about 15 minutes.

Watchpoint The mixture should hold its shape – if too much milk is added the cakes lose their rock-like appearance.

Freeze and thaw as for English madeleines.

Rich almond cake

4 oz butter
5 oz caster sugar
3 eggs
3 oz ground almonds
1½ oz plain flour
2–3 drops of almond essence

Deep 7-inch diameter sandwich tin

Method

Grease and flour sandwich tin, cover base with disc of greaseproof paper; set the oven at 350°F or Mark 4.

Soften butter with a wooden spoon in a bowl, add the sugar a tablespoon at a time, and beat thoroughly until mixture is soft and light. Add the eggs, one at a time, adding one-third of the almonds with each egg. Beat well. Fold in the flour and almond essence with a metal spoon and turn cake mixture into the prepared tin.

Bake in pre-set oven for 45–50 minutes until cake is cooked. (Test by inserting a thin skewer; it should come out clean.) When cooked, the cake should also shrink very slightly from the sides of the tin.

To turn out, have ready two wire cooling racks, put a clean folded tea towel or single thickness of absorbent paper on one of them. Loosen the sides of the cake with a round-bladed knife, place the rack with the towel or paper on top of the cake (towel next to it) and turn over; remove the tin and disc of paper from the base.

Place second rack on top of cake base and carefully and quickly turn it over again. This prevents the rack marking top of cake.

Freeze as for Victoria sandwich (see page 106).

To use, thaw at room temperature for 1–2 hours, dust with caster sugar and serve with fruit compote.

Strawberry walnut cream cake

3 eggs
scant 4 oz sugar
3 oz plain flour
2 oz walnuts (coarsely chopped)
2 tablespoons coffee essence

For filling
8–10 fl oz double cream (lightly whipped)
1 lb strawberries

Deep 8-inch diameter cake tin

Method

Prepare the tin by greasing and dusting out with caster sugar and flour. Set the oven at 350°F or Mark 4.

To prepare the cake mixture: whisk the eggs and sugar over heat as for a whisked sponge, or combine using an electric mixer. When really thick and mousse-like, take off the heat and continue to whisk for a further minute. Sift the flour and fold it in with the walnuts and coffee essence. Turn the cake mixture into the tin and bake in the pre-set oven for 40–45 minutes.

Turn cake out to cool and when cold split into three and layer with two-thirds of the lightly whipped cream mixed with the sliced strawberries. (Reserve a few strawberries for decoration.) The filling should be really lavish. Spread the rest of the cream over the top of the cake and decorate with the reserved strawberries.

Pack in a rigid container or cardboard box and freeze.

To use, thaw in the refrigerator for 6–7 hours.

Christmas chocolate log

For swiss roll
1 oz plain flour
1 dessertspoon cocoa
pinch of salt
3 eggs
large pinch of cream of tartar
4 oz caster sugar
2–3 drops of vanilla essence

For butter cream
3 oz caster sugar
3 egg yolks
6 oz unsalted butter
melted chocolate, for flavouring

For meringue mushrooms
1 egg white
2 oz caster sugar
little icing sugar
little chocolate (grated)

For decoration
6 oz boiled marzipan
little instant coffee
green colouring
little icing sugar

Swiss roll tin, or paper case, 8 inches by 12 inches; forcing bag, 6-cut vegetable rose pipe and ¼-inch plain pipe

Method

Set oven at 325°F or Mark 3; grease and flour swiss roll tin or paper case.

Sift the flour well with the cocoa and salt. Separate the eggs and whisk the whites with the cream of tartar until stiff; then gradually beat in half the sugar. Continue whisking until the mixture looks very glossy and will stand in peaks.

Cream the egg yolks until thick, then beat in the remaining sugar and add the vanilla essence. Stir the flour into the yolks and pour this mixture over the whites. Using a metal spoon, cut and fold carefully until thoroughly blended.

Turn the mixture into the prepared tin or paper case, bake in the pre-set oven for 20–25 minutes.

Meanwhile, make the butter cream. Dissolve the sugar in 3–4 fl oz water over gentle heat, then boil steadily to 216–218°F. Pour the syrup while still hot on to the egg yolks and whisk until a thick mousse is formed. Cream the butter until soft and beat in the eggs and sugar mousse a little at a time. Flavour with melted chocolate.

When the cake is cooked, turn it at once on to a sugared tea towel, trim the edges and roll it up, with the tea towel inside the cake. When cool, unroll the cake carefully and spread with chocolate butter cream. Roll up again.

To make meringue mushrooms: turn down oven to 275°F or Mark 1. Whisk the egg white until stiff, whisk in 2 teaspoons of the sugar and then fold in the rest with a metal spoon. Using a forcing bag and a plain ¼-inch diameter pipe, pipe out several small mushroom caps and stalks on an oiled and floured baking sheet. Dust the caps with icing sugar and grated chocolate and bake in the pre-set oven for about 45 minutes. Cool. Make a small dent in the underside of each cap and fix the stalks in place with a little butter cream.

Take two-thirds of the boiled marzipan and colour with a little instant coffee. Make into small rounds to represent knots in the log and to form the two ends of the log. Colour the remaining marzipan green and roll it out. Shape and cut it to represent a trail of ivy.

Spread the swiss roll with a light coating of butter cream. Then place the two marzipan ends on the log. Pipe the remaining butter cream on to the log, using a 6-cut vegetable rose pipe, placing the marzipan knots along it. Put the trail of ivy over the log. Arrange the meringue mushrooms arround the side and sprinkle with icing sugar to look like snow.

Pack the log carefully in a cardboard box and freeze. To use, remove the lid of the box and thaw at room temperature for about $2\frac{3}{4}$ hours.

Forming coloured almond paste into knots and ends for the Christmas log

Gâteau moka aux amandes

3 oz plain flour
pinch of salt
3 eggs
4½ oz caster sugar

To finish
butter cream (see page 110)
coffee essence (to flavour)
3 oz almonds (blanched, split and shredded)
icing sugar

9½-inch diameter layer cake tin

Method

Set the oven at 370°F or Mark 4–5. Grease and flour the tin.

Sift the flour with salt. Break the eggs into a bowl, add the sugar gradually and then whisk over a pan of hot water until the mixture is thick and white. Remove bowl from the heat and continue beating until the bowl is cold. Fold the flour into the mixture, using a tablespoon. Pour the mixture at once into the tin and bake it in pre-set oven for 25–30 minutes. Turn the cake on to a wire rack to cool.

Have ready the butter cream well flavoured with the coffee essence. Bake almonds to a light golden-brown.

Split the cake and sandwich with a thin layer of the coffee butter cream. Reshape and spread the top and sides with more of the same cream, press the browned almonds all over the cake, dredge lightly with icing sugar and then decorate round the top edge with rosettes of butter cream.

Pack in a rigid plastic container or cardboard box and freeze.

To use, thaw at room temperature for 4–5 hours.

Gâteau flamande

4 oz quantity of French flan pastry (see page 125)
2 oz crystallised cherries
2–3 tablespoons kirsch

For frangipane
4 oz butter
4 oz caster sugar
2 eggs
4 oz ground almonds
1 oz plain flour (sifted)
1 tablespoon kirsch

To decorate
2 oz flaked almonds
4 tablespoons thick glacé icing

8-inch diameter flan ring

Method

Make up the pastry, chill and then line into flan ring. Slice cherries, reserving a few for decoration and macerate in the kirsch. Set oven at 375°F or Mark 5.

Soften the butter, add the sugar and beat until light and fluffy. Beat in the eggs a little at a time, then stir in the almonds, flour and kirsch.

Place the cherries at the bottom of the flan, cover with frangipane and place the almonds on top. Bake gâteau in pre-set oven for about 45 minutes. When cool, brush gâteau with glacé icing and decorate with reserved cherries.

Pack in a rigid plastic container or cardboard box and freeeze.

To use, thaw at room temperature for 4–5 hours.

Gâteau au chocolat

2¼ oz plain flour
pinch of salt
2 oz chocolate (unsweetened)
about 2½ fl oz water
3 eggs
4½ oz caster sugar
2 oz plain chocolate
6 oz butter cream (see page 110)
chocolate caraque
a little icing sugar (to decorate)

9½-inch diameter layer cake tin

Chocolate caraque
Melt 3 oz grated plain chocolate or chocolate couverture (cooking chocolate) on a plate over hot water and work with a palette knife until smooth. Spread thinly on a marble slab or laminated surface and leave until nearly set. With a sharp long knife, using a slight sawing movement and holding knife almost upright, shave off long scrolls or flakes.

Method
Set oven at 350°F or Mark 4. Grease and flour the cake tin.
  Sift the flour with the salt. Grate or slice the unsweetened chocolate, melt it in the water over gentle heat until it is a thick cream, then set it aside to cool.
  Whisk the eggs and sugar together over gentle heat until thick and mousse-like, remove the bowl from the heat and continue whisking until the mixture is cold. Fold the flour into the mixture, then add the melted chocolate. Turn the mixture into the prepared tin and bake in pre-set oven for about 35 minutes.
  While the cake is cooking, melt the plain chocolate on a plate over a pan of hot water; when it is quite smooth, beat it into the butter cream.
  When cake is cool, split it in two halves and sandwich together with a thin layer of the chocolate butter cream. Reshape the cake, spread the top and sides with the same cream and press chocolate caraque over and around it, then sprinkle with icing sugar.
  Pack in a rigid plastic container or cardboard box and freeze.
  To use, thaw at room temperature for 4–5 hours.

Apple pie

8 oz shortcrust pastry (see page 118)
caster sugar (for dusting)

For filling
$1\frac{1}{4}$ lb cooking or mildly acid dessert apples
1 strip of lemon rind
2–3 rounded tablespoons sugar (brown or white)
little grated lemon rind, or 1–2 cloves (optional)

8-inch diameter pie dish

This is one of the oldest of English dishes and delicious when properly made. Use a fair-sized dish that holds plenty of fruit. Apples may be cookers, preferably a variety that will retain shape when cooked, or a mildly acid dessert apple such as Cox's orange pippin. A Blenheim orange apple, though not easy to find these days, makes an excellent pie, and combines both the qualities of dessert and cooking apples.

Method

To get the most flavour, peel, quarter and core the apples, keeping them in a covered bowl (not in water) while making juice from the cores and peel. Put these last in a pan with a strip of lemon rind, barely cover with water and simmer for 15–20 minutes. Then strain. (Water can, of course, be used in place of the juice made from the peel.)

While the juice is simmering, prepare a good shortcrust pastry. Put in the refrigerator to chill, or set aside in a cool place for about 30 minutes.

Line the pie dish with foil. Cut apple quarters into 2–3 pieces, pack into pie dish.

Watchpoint Do not slice the apples too thinly or the juice will run too quickly and may render the slices tough and tasteless.

Layer these slices with 2–3 tablespoons of sugar, according to the acidity of the fruit. Add, too, a little grated lemon rind or 1–2 cloves. Dome the fruit slightly above the edge of the dish (this is sufficient to prevent the pastry top from falling in, so there is no need to use a pie funnel). Pour in enough of the strained apple juice (or water) to fill it half full.

Take up the pastry and roll out to about $\frac{1}{4}$-inch thick, cut a strip or two from the sides and lay these on dampened edge of the pie dish. Press down and brush them with water. Take up the rest of the pastry on the rolling pin and lay it over the pie. Press down the edge, then lift the dish up on one hand and cut the excess pastry away, holding the knife slantwise towards the bottom of the dish to get a slightly overhanging edge. Pinch or scallop round the edge with your fingers. Freeze uncovered. When solid, remove the pie dish, completely cover with foil and over-wrap with polythene.

To use, remove the wrappings and place the frozen pie in the pie dish. Bake in the oven, preset at 425°F or Mark 7, for about 45 minutes. Dust the crust with caster sugar after baking.

Cherry pie

8 oz American pie pastry

For filling
2½ cups cherries (fresh or canned, stewed and stoned)
8 fl oz cherry juice
2 rounded tablespoons caster sugar
1 tablespoon melted butter
1 tablespoon fine tapioca, or sago
2 drops of almond essence
1 egg white (for glazing)
cream (optional)

7–8 inch diameter pie plate about 2½ inches deep

Method

Make American pie pastry and set aside to chill. Either fresh or canned Morello or red cherries (ie. not too sweet) are best for this pie. If fresh cherries are used, stone and cook in a little sugar syrup, and drain well.

Mix all the ingredients for the filling together and allow to stand for 15 minutes. Lay the pastry on to the foil-lined pie plate, pour in the fruit mixture and cover with the remaining pastry. Freeze uncovered. When solid, remove the pie plate, cover completely with foil and overwrap with polythene.

To use, remove the wrappings and place the frozen pie in the pie plate. Bake in the oven, pre-set at 425°F or Mark 7, for about 45 minutes. When it starts to thaw, slit the top crust in the centre. Dust with caster sugar and serve with whipped cream.

American pie pastry

Pastry for the American covered pie is slightly different from shortcrust both in ingredients and method. Most recipes for American shortcrust have a high proportion of fat to flour, and usually need more liquid for binding. This is because American and Canadian flour is milled from hard wheat which is very high in gluten (the major part of the protein content of wheat flour, which gives it its elasticity), and consequently absorbs more liquid.

The following recipe is an anglicised version, but has the same short, melt-in-the-mouth texture. As the texture is very short, the pastry is not easy to handle once cooked, so serve the pie in the dish in which it is baked (a round, shallow tin or dish—pie plate—about 2–2½ inches deep). The pastry is lined into the pie plate, fruit or other mixture is poured on top, and the pie is then covered with a lid of pastry.

Basic recipe

8 oz self-raising flour
5 oz lard, or shortening
pinch of salt
2 tablespoons cold water

Method

Place the lard or shortening in a bowl, add a good pinch of salt and the water, and cream ingredients together. Sift the flour over the softened fat and, using a round-bladed knife, cut the fat into the flour and mix to a rough dough. Chill for 30 minutes.

Turn the dough on to a floured board, knead lightly and then use for covered fruit pies.

Pumpkin pie Cordon Bleu

8 oz quantity of American pie pastry (as previous recipe)

For filling
½ lb mashed cooked pumpkin
4 oz soft brown sugar
1 tablespoon thick honey
grated rind and juice of 1 lemon
grated rind and juice of 1 orange
2 eggs (well beaten)

To finish
¼ pint double cream
1 teaspoon caster sugar
¼ teaspoon grated nutmeg
2 oz walnuts (roughly chopped)

Deep 8-inch diameter pie plate

Method

Line the pie plate with pastry. To prepare filling: mix the sugar, honey and fruit juices together, add the beaten eggs and the grated fruit rinds, then stir in the mashed pumpkin. Pour the mixture into the prepared pastry case. Freeze uncovered. When solid, remove the pie plate, cover with foil and overwrap with polythene.

To use, remove the wrappings, place the frozen pie in the pie plate. Bake in the oven, preset at 425°F or Mark 7, for 50–60 minutes, or until a knife inserted in filling comes out clean. Allow to cool.

Whip the cream, sweeten it with the sugar and add the nutmeg. Just before serving, cover the pie with the cream and sprinkle with chopped nuts.

Shortcrust pastry

8 oz plain flour
pinch of salt
4–6 oz butter, margarine, lard or shortening (one of the commercially prepared fats), or a mixture of any two
3–4 tablespoons cold water

Method

Sift the flour with a pinch of salt into a mixing bowl. Cut the fat into the flour with a round-bladed knife and, as soon as the pieces are well coated with flour, rub in with the fingertips until the mixture looks like fine breadcrumbs.

Make a well in the centre, add the water (reserving about 1 tablespoon) and mix quickly with a knife. Press together with the fingers, adding the extra water, if necessary, to give a firm dough.

Turn on to a floured board, knead pastry lightly until smooth. Wrap in polythene and freeze.

To use, thaw at room temperature for 1 hour. Roll out and use as required.

Alternatively, pastry may be frozen at the rubbed-in stage, ready for making into a dough when required.

Rich shortcrust pastry

8 oz plain flour
pinch of salt
6 oz butter
1 rounded dessertspoon caster sugar (for sweet pastry)
1 egg yolk
2–3 tablespoons cold water

Method

Sift the flour with a pinch of salt into a mixing bowl. Drop in the butter and cut it into the flour until the small pieces are well coated. Then rub them in with the fingertips until the mixture looks like fine breadcrumbs. Stir in the sugar, mix egg yolk with water, tip into the fat and flour and mix quickly with a palette knife to a firm dough.

Turn on to a floured board and knead lightly until smooth.

Freeze as for shortcrust pastry.

Savoury shortcrust pastry

4 oz plain flour
salt and pepper
pinch of cayenne pepper
2 oz shortening
$\frac{1}{2}$ oz Parmesan cheese (grated)
$\frac{1}{2}$ egg yolk (mixed with 1 tablespoon water)

Use this pastry for canapés, small boat moulds and tartlet tins.

This quantity will make about twelve $1\frac{1}{4}$-inch diameter canapés, or fill 12–16 boat moulds or tartlet tins.

Method
Sift the flour with seasonings, rub the shortening into the flour until the mixture resembles breadcrumbs. Add the cheese and mix to a dough with the egg yolk and water.

Freeze as for shortcrust pastry.

Savoury almond pastry

4 oz plain flour
salt and pepper
pinch of cayenne pepper
2 oz shortening
$\frac{1}{2}$ oz almonds (ground)
$\frac{1}{2}$ egg yolk (mixed with 1 tablespoon water)

Method
Make as for savoury shortcrust (see left), adding the ground almonds after the shortening has been rubbed into the flour.

Freeze as for shortcrust pastry.

Rough puff pastry 1

8 oz plain flour
pinch of salt
6 oz firm butter, or margarine
$\frac{1}{4}$ pint ice-cold water (to mix)

The first of the two types of rough puff pastry is a quicker and less fussy one, although the same ingredients are used in both types. You can use either type in recipes but the second is likely to be a little lighter.

Method

Sift the flour with salt into a mixing bowl. Cut the fat in even-size pieces about the size of walnuts and drop into the flour. Mix quickly with the water (to prevent overworking dough so that it becomes starchy) and turn on to a lightly-floured board.

Complete the following action three times; roll to an oblong, fold in three and make a half-turn to bring the open edges in front of you so that the pastry has three turns in all. Chill for 10 minutes and give an extra roll and fold if it looks at all streaky.

Freeze as for shortcrust pastry.

Rough puff pastry 2

8 oz plain flour
pinch of salt
6 oz firm butter, or margarine
$\frac{1}{4}$ pint ice-cold water (to mix)

Method

Sift the flour with salt into a mixing bowl. Take 1 oz of fat and rub it into the flour. Mix to a firm but pliable dough with the water, knead lightly until smooth, then set in a cool place for 10–15 minutes.

Place the remaining fat between two pieces of greaseproof paper and beat to a flat cake with the rolling pin. This fat should be the same consistency as the dough.

Roll out this dough to a rectangle, place the flattened fat in the middle, fold like a parcel and turn over.

Complete the following action three times: roll out dough to an oblong, fold in three and make a half-turn to bring the open edge towards you so that the pastry has three turns in all.

Freeze as for shortcrust pastry.

Flaky pastry

8 oz plain flour
pinch of salt
3 oz butter
3 oz lard
¼ pint ice-cold water (to mix)

Method

Sift the flour with salt into a bowl. Divide the fats into four portions (two of butter, two of lard); rub one portion – either lard or butter – into the flour and mix to a firm dough with cold water. The amount of water varies with different flour but an average quantity for 8 oz flour is 4–5 fluid oz (about ¼ pint or 8–10 tablespoons); the finer the flour the more water it will absorb.

Knead the dough lightly until smooth, then roll out to an oblong. Put a second portion of fat (not the same kind as first portion rubbed in) in small pieces on to two-thirds of the dough. Fold in three, half turn the dough to bring the open edge towards you and roll out again to an oblong. Put on a third portion of fat in pieces, fold dough in three, wrap in a cloth or polythene bag and leave in a cool place for 15 minutes.

Roll out dough again, put on remaining fat in pieces, fold and roll as before. If pastry looks at all streaky, give one more turn and roll again.

Freeze as for shortcrust pastry.

1 *Mix fat and flour to a firm dough with a little ice-cold water*

2 *Knead dough, roll out to an oblong, dot second portion of fat over two-thirds of dough.*

3 *Roll out the dough again, put on the last portion of fat and roll out as before*

Puff pastry

Forming the dough
To have perfect results when making puff pastry, you must use the right kind of flour and fat, and always use ice-cold water for mixing.
It is also important to work in a very cool atmosphere. Never attempt to make puff pastry in very hot weather; it will become sticky and difficult to handle. Make it early in the morning (if possible before you have done any cooking), as a kitchen soon becomes warm and steamy.
Fat should be cool and firm. The best puff pastry for flavour and texture is made from butter; this should be of a firm consistency and slightly salted — such as English, Australian or New Zealand. Continental butters are too creamy in texture and result in a sticky pastry, difficult to handle.
If margarine has to be used, again use a firm variety (one that does not spread easily). The cheapest varieties of butter and margarine are the best for this purpose.

Flour should be 'strong', ie. a bread flour which has a high gluten content. It should also be well sifted and quite cool.
The flour is made into a firm dough with a little butter and the water. This preliminary mixing is most important as it is on this that the success of the pastry depends.
Add the lemon juice to approximately two-thirds of the given amount of water. Stir until a dough begins to form, then add remaining water. If water is added a little at a time, it will dry in the flour and the resulting dough will be tough. The finished dough should be firm yet pliable and have the consistency of butter, taking into account the different textures.
Knead the dough firmly — this and the presence of the lemon juice develops the gluten in the flour and means that the dough will stand the frequent rolling and folding necessary in the preparation of puff pastry.
The butter should be cool and firm, but not used straight from the refrigerator. If it is overhard (or not taken from the refrigerator early enough), put it between two pieces of damp greaseproof paper and beat it 2–3 times with the rolling pin. It is then ready to be rolled into the dough.

Rolling out the dough
The method of rolling is also important and differs slightly from the usual way. You roll shortcrust pastry to shape the dough; in puff pastry it is the rolling that actually makes it.
Always roll the dough away from you, keeping the pressure as even as possible. Many people are inclined to put more weight on the right or left hand, which pulls the dough to one side; keep it straight by applying even pressure all round.
Bring the rolling pin down smartly on to the dough and roll it forward with a strong, firm pressure in one direction only. Continue until just before the edge of the dough.
Watchpoint Never let the rolling pin run off the edge as the object is to keep the dough strictly rectangular in shape.

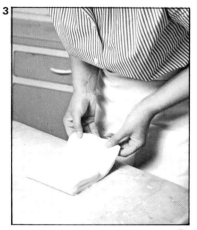

Lift the rolling pin and continue rolling forward in one direction, bringing it down at the point to which it was last rolled. In this way the whole area of the dough is rolled in even layers, $\frac{1}{2}$–$\frac{3}{4}$ inch thick.

Once rolled to an even rectangle, the dough is folded in three round the butter (see method overleaf). Graduate the

1 *After first rolling out of the pastry, butter is laid on the centre and sides turned in over it*

2 *Rolling pin is brought down lightly on to the pastry to flatten it before rolling out*

3 *The pastry is folded into three, ends to middle, like a parcel*

4 *After each rolling, pastry is always folded into three, the ends pulled to keep them rectangular*

Puff pastry continued

thickness of the following rollings, so that these subsequent ones are progressively thinner. You must avoid pushing butter through the dough, which might happen if it was rolled thinly in the beginning.

Watchpoint Do not turn the dough over; it should only be rolled on one side.

Each rolling is called a 'turn' and puff pastry usually has six turns with a 15-minute rest between every two. Before each turn the dough is folded in three (ends to middle) and the edges sealed with the side of the hand or the rolling pin to prevent the folds shifting when dough is rolled. The short period of rest is to remove any elasticity from the dough. If at the end of the rollings the dough is at all streaky (showing that the butter has not been rolled in completely), a seventh turn can be given.

Should fat begin to break through dough, stop at once. Dust dough with flour, brush off the surplus, and chill it for 10 minutes before continuing.

Basic puff pastry

8 oz plain flour
pinch of salt
8 oz butter
1 teaspoon lemon juice
scant $\frac{1}{4}$ pint water (ice cold)

This quantity will make a vol-au-vent for 4 people or 6–8 medium-size bouchées.

Method

Sift flour and salt into a bowl. Rub in a piece of butter the size of a walnut. Add lemon juice to water, make a well in centre of flour and pour in about two-thirds of the liquid. Mix with a palette, or round-bladed, knife. When the dough is beginning to form, add remaining water.

Turn out the dough on to a marble slab, a laminated-plastic work top, or a board, dusted with flour. Knead dough for 2–3 minutes, then roll out to a square about $\frac{1}{2}$–$\frac{3}{4}$ inch thick.

Beat butter, if necessary, to make it pliable and place in centre of dough. Fold this up over butter to enclose it completely (sides and ends over centre like a parcel). Wrap in a cloth or piece of greaseproof paper and put in the refrigerator for 10–15 minutes.

Flour slab or work top, put on dough, the join facing upwards, and bring rolling pin down on to dough 3–4 times to flatten it slightly.

Now roll out to a rectangle about $\frac{1}{2}$–$\frac{3}{4}$ inch thick. Fold into three, ends to middle, as accurately as possible, if necessary pulling the ends to keep them rectangular. Seal the edges with your hand or rolling pin and turn pastry half round to bring the edge towards you. Roll out again and fold in three (keep a note of the 'turns' given). Set pastry aside in refrigerator for 15 minutes.

Repeat this process, giving a total of 6 turns with three 15-minute rests after each two turns.

Freeze as for shortcrust pastry.

French flan pastry

4 oz plain flour
2 oz butter
2 oz caster sugar
2–3 drops of vanilla essence
2 egg yolks

Note: 2 oz vanilla sugar may be used instead of caster sugar and vanilla essence.

Method

Sift the flour with a pinch of salt on to a marble slab or pastry board, make a well in the centre and in it place the butter, sugar, vanilla essence and egg yolks. Using the fingertips of one hand only, pinch and work these last three ingredients together until well blended. Then draw in the flour, knead lightly until smooth.
Freeze as for shortcrust pastry.

Genoese pastry

$4\frac{1}{4}$ oz plain flour
pinch of salt
2 oz butter
4 eggs
$4\frac{1}{4}$ oz caster sugar

$8\frac{1}{2}$–9 inch diameter moule à manqué

Method

Set the oven at 350–375°F or Mark 4–5; grease mould, line the bottom only with a disc of greaseproof paper to fit exactly, grease again, dust with caster sugar, then flour.
Sift the flour 2 or 3 times with the salt. Warm the butter gently until soft and pourable, taking great care not to make it hot or oily. Have ready a large saucepan half full of boiling water over which the mixing bowl will rest comfortably without touching the water.
Break the eggs into the bowl and beat in the sugar gradually. Remove the saucepan from the heat, place the bowl on top and whisk the eggs and sugar until thick and mousse-like. This will take 7–8 minutes and the mixture will increase in volume and lighten in colour. Remove the bowl from the heat and continue whisking for 5 minutes until the mixture is cold. Using a metal spoon, fold in two-thirds of the flour, then the butter, followed by the remaining flour.
Turn the mixture into the prepared mould and bake in the pre-set oven for 30–35 minutes. Cool on a wire rack. Freeze quickly and pack in a rigid container.
To use, thaw at room temperature for 1–2 hours.

Flans

Flans are open tarts where the pastry (shortcrust, or French flan pastry known as pâte sucrée) is rolled out and lined on to a flan ring laid on a baking sheet. The flan is filled with a cake or savoury mixture, pastry cream, or more usually with fruit.

There are two types of flan ring, the 1-inch deep British one and the true French kind which is barely $\frac{3}{4}$ inch deep. This latter ring is the correct one for all fruit flans and can be found in specialist shops. The deeper ring is good for savoury flans, where a generous amount of filling is used. A flan can be made in a loose-bottomed sandwich tin, but to avoid breaking pastry, or a burn, leave it to cool before removing from the tin.

Apples, gooseberries and stone fruit may be cooked in the raw pastry flan. Other fruits, such as raspberries or poached fruit, are arranged in the pre-cooked flan (this is known as baking blind).

Once filled all fruit flans are glazed, either with a thickened fruit juice (particularly if cooked or canned fruit is used), or a jam or jelly glaze. For serving use a flat plate or board, not a shallow dish.

Flan cases may be frozen filled or empty. If freezing empty cases, wrap them very carefully since cooked pastry is fragile. To use, thaw at room temperature for 1 hour and fill as required. If required hot, re-heat briefly in the oven. Filled cases will require 2–4 hours to thaw, depending on size.

Lining a flan ring

1   Have ready the pastry, well chilled. Set the flan ring on a baking sheet, preferably without edges, for easy removal of the flan. Roll out the pastry to a thickness of $\frac{1}{4}$–$\frac{1}{2}$ inch, according to the recipe, and to a diameter about $1\frac{1}{2}$ inches bigger than the flan ring. Lift the pastry up on the rolling pin and lay over the flan ring, quickly easing it down into the ring.

2   Take a small ball of the dough, dip in flour and press the pastry into the ring, especially round the bottom edge.

3   Now bend back the top edge and roll off excess pastry with the rolling pin.

4   Pinch round the edge with the side of the forefinger and thumb, then push the dough (with the fingers) up the side from the bottom of the ring to increase the height of the edge. Prick the pastry base of the flan several times with a fork. Then fill with raw fruit.

Baking blind

1   A flan case should be pre-cooked before filling with soft or cooked fruit. Once the flan ring is lined with pastry, chill for about 30 minutes to ensure the dough is well set.

2   Now line the pastry with crumpled greaseproof paper, pressing it well into the dough at the bottom edge and sides.

3   Three-parts fill the flan with uncooked rice or beans (to hold

the shape) and put into the oven to bake. An 8-inch diameter flan ring holding a 6–8 oz quantity of pastry should cook for about 26 minutes in an oven at 400°F or Mark 6.

4 After about 20 minutes of the cooking time take flan out of the oven and carefully remove the paper and rice, or beans.

(Beans may be used many times over for baking blind.) Replace the flan in the oven to complete cooking. The ring itself can either be taken off with the paper and rice, or removed after cooking. Once cooked, slide the flan on to a wire rack and then leave to cool.

Lift pastry up on rolling pin, lay over the flan ring, quickly easing it down into the ring

Press pastry into ring with ball of floured dough. Bend back top edge, roll off any excess pastry

Pinch round edge with forefinger and thumb; push up sides from bottom of ring to raise edge

Before baking blind, line with grease-proof paper, then three parts fill with beans or rice

White sauce

¾ oz butter
1 rounded tablespoon plain flour
½ pint milk
salt and pepper

A white sauce is quick and easy, made in exactly the same way and with same proportions as béchamel, but the milk is not flavoured. It can be used as the base for cheese, onion or other sauces with pronounced flavour, but béchamel is better for mushroom and egg sauces.

Method
Melt the butter in a small pan, remove from heat and stir in the flour. Blend in half the milk, then stir in the rest. Stir this over moderate heat until boiling, then boil gently for 1–2 minutes. Season to taste.
 Freeze in a rigid container.
 To use, turn into a pan and heat gently, stirring all the time. Bring to the boil and adjust seasoning.

Béchamel sauce

½ pint milk
1 slice of onion
1 small bayleaf
6 peppercorns
1 blade of mace

For roux
¾ oz butter
1 rounded tablespoon plain flour
salt and pepper

Made on a white roux with flavoured milk added, béchamel can be used as a base for mornay (cheese), soubise (onion), mushroom or egg sauces. Proportions of ingredients may vary in these derivative sauces according to consistency required.

Method
Pour milk into a saucepan, add the flavourings, cover pan and infuse on gentle heat for 5–7 minutes. Strain milk and set it aside. Rinse and wipe out the pan and melt the butter in it. To give a white roux remove from heat before stirring in the flour. The roux must be soft and semi-liquid.
 Pour on half of milk through a strainer and blend until smooth using a wooden spoon, then add rest of milk. Season lightly, return to a slow to moderate heat and stir until boiling. Boil for no longer than 2 minutes.
 Freeze and thaw as for White Sauce.

Velouté sauce

¾ oz butter
1 rounded tablespoon plain flour
⅓–½ pint stock
2½ fl oz top of milk
salt and pepper
squeeze of lemon juice

For liaison (optional)
1 egg yolk (lightly beaten)
2 tablespoons cream

This sauce is made with a blond roux, to which liquid is added. This is well-flavoured stock (made from veal, chicken or fish bones, according to dish with which sauce is being served), or liquid in which food was simmered or poached.

Velouté sauces are a base for others, such as caper, mustard, parsley, poulette or suprême.

Method

Melt butter in a saucepan, stir in flour and cook for about 5 seconds (blond roux). When roux is colour of pale straw, draw pan aside and cool slightly before pouring on stock.

Blend, return to heat and stir until thick. Add top of milk, season and bring to boil. Cook 4–5 minutes when sauce should be a syrupy consistency. Freeze at this point. If using a liaison, add it after thawing. Prepare by mixing egg yolk and cream together and then stir into sauce. Add lemon juice. Remove pan from heat.

Watchpoint Be careful not to let sauce boil after liaison has been added, otherwise the mixture will curdle.

Hollandaise sauce

4 tablespoons white wine vinegar
6 peppercorns
1 blade mace
1 slice of onion
1 small bayleaf
3 egg yolks
5 oz butter (unsalted)
salt and pepper
1–2 tablespoons single cream, or top of milk
squeeze of lemon juice (optional)

Method

Put the vinegar into a small pan with the spices, onion and bayleaf. Boil this until reduced to a scant tablespoon, then set aside.

Cream egg yolks in a bowl with a good nut of butter and a pinch of salt. Strain on the vinegar mixture, set the bowl on a pan of boiling water, turn off heat and add remaining butter in small pieces, stirring vigorously all the time.

Watchpoint When adding butter, it should be slightly soft, not straight from refrigerator.

When all the butter has been added and the sauce is thick, taste for seasoning and add the cream or milk and lemon juice. The sauce should be pleasantly sharp yet bland, and should have consistency of thick cream.

Freeze in a rigid container. Thaw very gently, preferably in a double saucepan.

Sauce blanche au beurre

2 oz butter
1 tablespoon plain flour
½ pint water (boiling)
salt and pepper
good squeeze of lemon juice

Method

Melt a good ½ oz of butter in a pan, stir in the flour off the heat and when smooth pour on all the boiling water, stirring or whisking briskly all the time.

Now add remaining butter in small pieces, stirring it well in. Season and add lemon juice.

Watchpoint If the water is really boiling it will cook flour. On no account bring sauce to the boil as this will give it an unpleasant gluey taste.

Freeze in a rigid container and thaw very gently, in a double saucepan.

Basic brown (demi-glace) sauce

3 tablespoons salad oil
1 small onion (finely diced)
1 small carrot (finely diced)
½ stick of celery (finely diced)
1 rounded tablespoon plain flour
1 teaspoon tomato purée
1 tablespoon mushroom peelings (chopped), or 1 mushroom
1 pint well-flavoured brown stock
bouquet garni
salt and pepper

Method

Heat a saucepan, put in the oil and then add diced vegetables (of which there should be no more than 3 tablespoons in all). Lower heat and cook gently until vegetables are on point of changing colour; an indication of this is when they shrink slightly.

Mix in the flour and brown it slowly, stirring occasionally with a metal spoon and scraping the flour well from the bottom of the pan. When it is a good colour draw pan aside, cool a

little, add tomato purée and chopped peelings or mushroom, ¾ pint of cold stock, bouquet garni and seasonings.

Bring to the boil, partially cover pan and cook gently for about 35–40 minutes. Skim off any scum which rises to the surface during this time. Add half the reserved stock, bring again to boil and skim. Simmer for 5 minutes. Add rest of stock, bring to boil and skim again.

Watchpoint Addition of cold stock accelerates rising of scum and so helps to clear the sauce.

Cook for a further 5 minutes, then strain, pressing vegetables gently to extract the juice. Rinse out the pan and return sauce to it. Partially cover and continue to cook gently until syrupy in consistency. Divide between three or four small rigid containers for freezing. It is particularly useful to have this sauce available in small quantities, for adding to gravies. To use, turn into a small pan and heat gently, stirring; bring to the boil and use as required.

Brown bone stock

3 lb beef bones (or mixed beef/veal)
2 onions (quartered)
2 carrots (quartered)
1 stick of celery
large bouquet garni
6 peppercorns
3–4 quarts water
salt

6-quart capacity saucepan, or small fish kettle

Method

Wipe bones but do not wash unless unavoidable. Put into a very large pan. Set on gentle heat and leave bones to fry gently for 15–20 minutes. Enough fat will come out from the marrow so do not add any.

After 10 minutes add the vegetables, having sliced the celery into 3–4 pieces.

When bones and vegetables are just coloured, add herbs, peppercorns and the water, which should come up two-thirds above level of ingredients. Bring slowly to the boil, skimming occasionally, then half cover pan to allow reduction to take place and simmer 4–5 hours, or until stock tastes strong and good.

Strain off and use bones again for a second boiling. Although this second stock will not be so strong as the first, it is good for soups and gravies. Use the first stock for brown sauces, sautés, casseroles, or where a jellied stock is required. For a strong beef broth, add 1 lb shin of beef.

Freeze in ½ pint quantities, in rigid containers. To use, turn into a saucepan and heat gently. When thawed, bring to boiling point before using.

Chicken stock

This should ideally be made from the giblets (neck, gizzard, heart and feet, if available), but never the liver which imparts a bitter flavour. This is better kept for making pâté, or sautéd and used as a savoury. Dry fry the giblets with an onion, washed but not peeled, and cut in half. To dry fry, use a thick pan with a lid, with barely enough fat to cover the bottom. Allow the pan to get very hot before putting in the giblets and onion, cook on full heat until lightly coloured. Remove pan from heat before covering with 2 pints of cold water. Add a large pinch of salt, a few peppercorns and a bouquet garni (bay leaf, thyme, parsley) and simmer gently for 1–2 hours. Alternatively, make the stock when you cook the chicken by putting the giblets in the roasting tin around the chicken with the onion and herbs, and use the measured quantity of water. Freeze as for brown stock.

White bone stock

This stock forms a basis for cream sauces, white stews, etc. It is made in the same way as brown bone stock, except that bones and vegetables are not browned before the water is added, and veal bones are used. Do not add the vegetables until the bones have come to the boil and the fat has been skimmed off the liquid. Freeze as for brown stock.

Jellied stock

This is invaluable for adding to the roasting tin to make a good gravy or to give flavour to a stew or ragoût. Use a veal bone, chicken giblets or a pig's trotter, and simmer carefully so that the stock stays clear. Strain it and allow to get cold before taking off the fat and storing. Freeze as for brown stock.

Vegetable stock

1 lb carrots
1 lb onions
$\frac{1}{2}$ head of celery
$\frac{1}{2}$ oz butter
3–4 peppercorns
1 teaspoon tomato purée
2 quarts water
salt

Method
Quarter vegetables, brown lightly in the butter in a large pan. Add peppercorns, tomato purée, water and salt. Bring to boil, cover pan and simmer 2 hours or until the stock has a good flavour. Freeze as for brown stock.

Mixed stock

If you want a really clear stock, the only way to make it is to use raw bones. If you are using cooked ones as well, it helps to add these after the stock has come to the boil, although it is better not to mix raw with cooked bones if the stock is to be kept for any length of time.

Any trimmings or leftovers in the way of meat can go into your regular stockpot: chicken carcasses and giblets (but not the liver); bacon rinds; or a ham or bacon bone. This last is often given away and it makes excellent stock for a pea soup.

Add a plateful of cut-up root vegetables, a bouquet garni, 5–6 peppercorns, and pour in enough cold water to cover the ingredients by two-thirds. Salt very lightly, or not at all if there is a bacon bone in the pot. Bring slowly to the boil, skim, half-cover the pan and simmer $1\frac{1}{2}$–2 hours or longer depending on the quantity of stock being made. The liquid should reduce by about a third. Strain off and, when the stock is cold, skim well to remove any fat. Throw away the ingredients unless a fair amount of raw bones have been used, in which case more water can be added and a second boiling made.

Watchpoint Long slow simmering is essential for any meat stock. It should never be allowed to boil hard as this will result in a thick muddy-looking jelly instead of a semi-clear one. Freeze as for brown stock.

Mixed or ordinary household stock is general-purpose and is used for making gravies, simple soups, broths and sauces

Maître d'hôtel butter

8 oz unsalted butter
4 dessertspoons chopped parsley
lemon juice
salt and pepper

Method

Soften the butter on a plate with a palette knife, then add parsley, lemon juice and seasoning to taste.

Chill until manageable, then place the butter on a board, between two sheets of greaseproof paper and shape into a roll, about 1 inch in diameter. Wrap in foil and polythene and freeze.

To use, simply slice off as many portions as are required from the end of the roll, to serve with steaks, mixed grills and fish. Return the rest to the freezer.

Orange butter

8 oz unsalted butter
grated rind of 2 oranges and $1\frac{1}{2}$ tablespoons juice
4 teaspoons tomato purée
salt and pepper

Method

Soften the butter on a plate with a palette knife, and then add other ingredients, seasoning to taste. Freeze as for maître d'hôtel butter.

Serve with lamb cutlets, steaks and fish.

Anchovy butter

8 oz unsalted butter
16 anchovy fillets (soaked in milk to remove excess salt)
black pepper (ground from mill)
anchovy essence

Method

Soften the butter on a plate with a palette knife and then crush or pound the anchovies, adding these to the butter with ground pepper and enough essence to strengthen the flavour and give a delicate pink colour. Freeze as for maître d'hôtel butter.

Serve with mutton, chops or cutlets, and fish.

Bolognese sauce 1

2 tablespoons oil, or 1 oz butter
4 oz chicken livers (approximately 3)
1 medium-size onion (sliced)
1 clove of garlic (chopped)
1 rounded dessertspoon plain flour
3 teaspoons tomato purée
$\frac{1}{2}$ pint of beef stock (or equivalent made from beef bouillon cube)
1 tablespoon Marsala, or brown sherry
salt and pepper
chopped parsley
Parmesan cheese (grated)

This sauce is best spooned over rather than mixed in with the spaghetti. Cook spaghetti, finish with oil or butter, turn into serving dish and spoon the sauce in a band over the top. Serve grated Parmesan cheese separately. Bolognese sauce should be made with chicken livers but lamb's liver can be substituted.

Method

Heat oil in shallow saucepan put in the livers, sauté for 3–4 minutes until 'seized' and nicely brown. Take out, add the onion and garlic, sauté until turning colour, then stir in the flour, add purée, stock and Marsala or sherry. Stir until boiling. Simmer for 10 minutes, then add liver, coarsely chopped.

Continue to simmer until thick and syrupy for a further 7–10 minutes Cool, turn into a rigid container and freeze.

To use, turn into a saucepan and heat gently, stirring from time to time. Bring to boiling point and simmer for 20–25 minutes to heat thoroughly.

Adjust seasoning and spoon sauce over the spaghetti. Sprinkle well with chopped parsley.

Bolognese sauce 2

2–3 tablespoons oil
1 medium-size onion (chopped)
2 oz mushrooms (sliced)
$\frac{1}{4}$ lb raw beef (finely minced)
1 rounded dessertspoon tomato purée
pinch of dried oregano
1 clove of garlic (crushed with salt)
$\frac{1}{4}$ of a green pepper (chopped) – optional
$7\frac{1}{2}$ fl oz stock

Method

Soften onion in the oil, add the mushrooms, the mince, purée, oregano, crushed clove of garlic and green pepper if used.

Fry mixture for a few minutes, then stir in the stock. Season, cover and simmer for 25–30 minutes or until meat is tender. Freeze and use as left.

Napolitana sauce

1 oz butter, or 2 tablespoons oil
1 medium-size onion (thinly sliced)
1 dessertspoon plain flour
1 wineglass stock
1 lb ripe tomatoes (skinned, the stalk cut out and the tomatoes lightly squeezed to remove seeds)
1 clove of garlic (crushed with salt)
1 teaspoon tomato conserve, or purée
1 bayleaf
pinch of sugar
salt
pepper (ground from mill)

Method

Melt butter or oil in a shallow saucepan, add onion, fry gently for 3–4 minutes, then stir in flour and add stock. Bring to the boil. Slice tomatoes and add with the garlic, conserve or purée, bayleaf, sugar, salt and pepper.

Simmer for 25–30 minutes or until well reduced to a thick rich pulp. Remove bayleaf.

Cool, turn into a rigid container and freeze.

To use, turn out into a saucepan. Heat gently, stirring from time to time. Bring to the boil and simmer for 10–15 minutes to heat thoroughly. Meanwhile, cook the pasta and place in a hot serving dish. Spoon the sauce over the top or mix with the pasta.

Appendix

Notes and basic recipes

Almonds

Buy almonds with their skins on. This way they retain their oil better. Blanching to remove the skins gives extra juiciness.

To blanch almonds: pour boiling water over the shelled nuts, cover the pan and leave until cool. Then the skins can be easily removed (test one with finger and thumb). Drain, rinse in cold water; press the skins off with the fingers. Rinse, dry thoroughly.

To grind almonds: first blanch and skin, chop and pound to a paste (use a pestle and mortar, or a grinder, or the butt end of a rolling pin). Home-prepared ground almonds taste much better than the ready ground variety.

Aspic jelly

2½ fl oz sherry
2½ fl oz white wine
2 oz gelatine
1¾ pints cold stock
1 teaspoon wine vinegar
2 egg whites

Add wines to gelatine and set aside. Pour cold stock into scalded pan, add vinegar. Whisk egg whites to a froth, add them to the pan, set over moderate heat and whisk backwards and downwards (the reverse of the usual whisking movement) until the stock is hot. Then add gelatine, which by now will have absorbed the wine, and continue whisking steadily until boiling point is reached.

Stop whisking and allow liquid to rise to the top of the pan; turn off heat or draw pan aside and leave to settle for about 5 minutes, then bring it again to the boil, draw pan aside once more and leave liquid to settle. At this point the liquid should look clear; if not, repeat the boiling-up process.

Filter the jelly through a scalded cloth or jelly bag. Allow to cool before using.

Breadcrumbs

To make fresh white crumbs: take a large loaf (the best type to use is a sandwich loaf) at least two days old. Cut off the crust and keep to one side. Break up bread into crumbs either by rubbing through a wire sieve or a Mouli sieve, or by working in an electric blender.

To make dried crumbs: spread crumbs on a sheet of paper laid on a baking tin and cover with another sheet of paper to keep off any dust. Leave to dry in a warm temperature — the plate rack, or warming drawer, or the top of the oven, or even the airing cupboard, is ideal. The crumbs may take a day or two to dry thoroughly, and they must be crisp before storing in a jar. To make them uniformly fine, sift them through a wire bowl strainer.

Court bouillon

(**2 pint quantity**)

2 pints water
1 carrot (sliced)
1 onion (stuck with a clove)
bouquet garni
6 peppercorns
2 tablespoons vinegar
salt

Method

Place all ingredients in a pan, salt lightly and bring to the boil. Cover the pan with a lid and simmer for 15-20 minutes. Strain before using.

Glacé icing

4-5 tablespoons granulated sugar
¼ pint water
8-12 oz icing sugar (finely sifted)
flavouring essence and colouring (as required)

Method

Make sugar syrup by dissolving sugar in the water in a small saucepan. Bring to the boil, and boil steadily for 10 minutes. Remove pan from the heat and when quite cold, add the icing sugar, 1 tablespoon at a time, and beat thoroughly with a wooden spatula. The icing should coat back of spoon and look very glossy. Warm the pan gently on a very low heat.

Watchpoint The pan must not get too hot. You should be able to touch the bottom with the palm of your hand.

Flavour and colour icing; spread over cake with a palette knife.

Marzipan, boiled

1 lb granulated sugar
6 fl oz and 1 teaspoon water
¾ lb ground almonds
2 egg whites (lightly beaten)
juice of ½ lemon
1 teaspoon orange flower water
3-4 tablespoons icing sugar

Method

Place the sugar and water in a saucepan and dissolve over gentle heat; bring it to the boil and cook steadily to 240°F. Remove the pan from the heat and beat the syrup until it looks a little cloudy, stir in the ground almonds, add the egg whites and cook over a gentle heat for 2-3 minutes. Add the flavourings and turn on to a marble slab or laminated surface dusted with icing sugar.

When the marzipan is cool, knead it until quite smooth. Colour and shape as required.

Redcurrant jelly

It is not possible to give a specific quantity of redcurrants as the recipe is governed by the amount of juice made, which is variable.

Method

Wash the fruit and, without removing from the stems, put in a 7 lb jam jar or stone crock. Cover and stand in a deep pan of hot water. Simmer on top of the stove or in the oven at 350°F or Mark 4, mashing the fruit a little from time to time until all the juice is extracted (about 1 hour).

Then turn fruit into a jelly bag or double linen strainer, and allow to drain undisturbed overnight over a basin.

Watchpoint To keep the jelly clear and sparkling, do not try to speed up the draining process by forcing juice through; this will only make the jelly cloudy.

Now measure juice. Allowing 1 lb lump or preserving sugar to each pint of juice, mix juice and sugar together, dissolving over slow heat. When dissolved, bring to the boil, boil hard for 3-5 minutes and skim with a wooden spoon. Test a little on a saucer; allow jelly to cool, tilt saucer and, if jelly is set, it will wrinkle. Put into jam jars, place small circles of greaseproof paper over jelly, label and cover with jam pot covers. Store in a dry larder until required.

Rice

Most people have their own favourite method of boiling rice. That recommended by Asians is to cook the rice in a small quantity of boiling water until this is absorbed, when rice is soft. The amount of water varies according to the quality of the rice. This method is good but can present problems. Really the simplest way is to cook the rice (about 2 oz washed rice per person) in plenty of boiling, well-salted water (3 quarts per 8 oz rice) for about 12 minutes. You can add a slice of lemon for flavour. Stir with a fork to prevent rice sticking while boiling, and watch that it does not over cook.

To stop rice cooking, either tip it quickly into a colander and drain, or pour $\frac{1}{2}$ cup cold water into the pan and then drain. Pour over a jug of hot water to wash away the remaining starch, making several holes through the rice with the handle of a wooden spoon to help it drain more quickly.

Tomatoes

To skin tomatoes: place them in a bowl, scald by pouring boiling water over them, count 12, then pour off the hot water and replace it with cold. The skin then comes off easily.

To remove seeds: slice off the top of each tomato and flick out seeds with the handle of a teaspoon, using the bowl of the spoon to detach the core.

Glossary

Bain-marie (au) To cook at a temperature just below boiling point in a bain-marie (a saucepan standing in a larger pan of simmering water). Used in the preparation of sauces, creams and food liable to spoil if cooked over direct heat. May be carried out in oven or on top of stove. A double saucepan gives a similar result. Sauces and other delicate dishes may be kept hot in a bain-marie at less than simmering heats.

Bouquet garni A bunch of herbs traditionally made up of 2–3 parsley stalks, a sprig of thyme and a bayleaf, tied together with string if used in liquids which are later strained. Otherwise herbs are tied in a piece of muslin for easy removal before serving the dish.

Croûton Small square or dice of fried bread or potato to accompany purée or cream soups.

Dégorger To remove impurities and strong flavours before cooking by **1.** Soaking food, eg. uncooked ham, in cold water for specified length of time. **2.** Sprinkling sliced vegetables, eg. cucumber, with salt, covering with a heavy plate, leaving up to 1 hour, and pressing out excess liquid with a weighted plate.

Kneaded butter A liaison of butter and flour worked together to form a paste (in the proportion 1 oz butter to $\frac{1}{2}$ oz flour). Added to the sauce in small pieces at the end of the cooking time. Useful when the exact quantity of liquid is not known.

Liaison Mixture for thickening/binding sauce/gravy/soup, eg. roux, egg yolks and cream, kneaded butter.

Macerate To soak/infuse, mostly fruit, in liqueur/syrup.

Marinate To soak raw meat/game/fish in cooked or raw spiced liquid (marinade) of wine, oil, herbs and vegetables for hours/days before cooking. This softens, tenderises and flavours and the marinade can be used in the final sauce. Use glass/stainless steel/enamel vessel to withstand effects of acid.

Mirepoix Basic preparation for flavouring braises and sauces. Diced vegetables, sweated (cooked gently for a few minutes in butter), to draw out flavour. Diced ham or bacon and bayleaf sometimes included.

Panade Basic thickening for fish/meat/vegetable creams. Often a thick béchamel sauce.

Roux Fat and flour liaison, the basis of all flour sauces. The fat is melted and the flour stirred in off the heat before the liquid is added.

Sauté To brown food in butter, or oil and butter. Sometimes cooking is completed in a 'small' sauce – ie. one made on the food in the sauté pan.

Scald 1. To plunge into boiling water for easy peeling. **2.** To heat a liquid, eg. milk, to just under boiling point.

Slake To mix arrowroot/cornflour with water before adding to a liquid for thickening.

Index

Almond(s), 138
 filling, 103
Apple
 filling, 103
 pie, 115
Asparagus, cream soup, 40
Aspic jelly, 138
Aubergines, stuffed, 48
Avocado pear and tomato ice, 45

Beef
 carbonade, 66
 en daube, 68
 sauté chasseur, 77
 Scotch collops, or mince, 74
 spiced 1 and 2, 90-91
Bean stew, Puchero, 73
Breadcrumbs, 138
Butter
 anchovy, 134
 maître d'hôtel, 134
 orange, 134

Cake(s), large
 Christmas chocolate log, 110
 gâteau au chocolat, 114
 gâteau flamande, 113
 gâteau moka aux amandes, 112
 rich almond, 108
 strawberry walnut cream, 109
 Victoria sandwich, 106
Cake(s), small
 English madelaines, 107
 rock, 107
Carbonade of beef, 66
Cauliflower au gratin, 97
Cherry pie, 116
Chestnut parfait, 100
Chicken
 broth, 37
 cockie-leekie soup, 37
 coq au vin, 78
 cuisses de poulet Xérès, 82
 curry, Madras, 88
 pâté, with calves liver, 46
 rolls, with tongue, 88
 spiced, 81
 suprêmes en fritot, 80
 terrine, 89
Chilli con carne, 72
Chocolate caraque, 114
Cockie-leekie soup, 37
Cod's roe pâté, 47
Consommé madrilène, 33
Cornish pasties, 74
Coq au vin, 78
Coquilles St. Jacques
 amoricaine, 44
Court bouillon, 138
Croissants, French, 105
Croûtons, 34

Cuisses de poulet Xérès, 82
Curry soup, iced, 42

Dolmades, 50
Duck, terrine, 89

Fish croquettes vert-pré, 61

Garam masala, 76
Glacé icing, 139

Ham pizza, 56
Hare
 jugged, 84
 terrine, 94

Koftas (meat balls), 76

Lamb
 goulash, 63
 koftas, 76
Liver, casserole, 72
Luting paste, 46

Marzipan, boiled, 139
Mulligatawny, 36

Ossi buchi, 69
Oxtail, braised, 71

Partridge, casserole, 87
Pastry(ies)
 American pie, 116
 baking blind, 126
 Danish, 102
 flaky, 121
 French flan, 125
 genoese, 125
 lining a flan ring, 126
 puff, 122
 rough puff 1 and 2, 120
 savoury almond, 119
 shortcrust, 118
  rich, 118
  savoury, 119
Pâté
 chicken and calves liver, 46
 cod's roe, 47
 country-style, 45
Peach(es)
 croissants with iced
  zabaione, 101
 gâteau with mousseline
  sauce, 100
 sliced, 98
Pie
 apple, 115
 cherry, 116
 pumpkin, Cordon Bleu, 117
 raised pork, 92
 steak and kidney, 64

Pizza, 53
  Cordon Bleu, 54
  ham, 56
  smoked haddock and mushroom, 56
  napolitana, 53
Pork
  brawn, 91
  pie, raised, 92
Potage bonne femme, 34
Potatoes
  chip, 96
  creamed, 96
Prawn bisque, chilled, 41
Puchero bean stew, 73
Pumpkin pie Cordon Bleu, 117

Quenelles de brochet, 62
Quiche lorraine, 52

Rabbit, casseroled with mustard, 86
Raspberries in strawberry cream, 98
Redcurrant Jelly, 139
Rice, 139

Salmon mousse, 57
Sauce
  basic brown (demi-glace), 130
  béchamel, 128
  blanche au beurre, 130
  bolognese 1 and 2, 135
  curry, 76
  hollandaise, 129
  mousseline, 100
  napolitana, 136
  tartare, 61
  tomato, 50
  velouté, 129
  white, 128
Scotch collops, or mince, 74

Seafood flan, 58
Sole georgette, 59
Soup
  asparagus, cream, 40
  chicken broth, 37
  cockie-leekie, 37
  consommé madrilène, 33
  curry, iced, 42
  mulligatawny, 36
  potage bonne femme, 34
  prawn bisque, chilled, 41
  vegetable bortsch, 38
Steak and kidney pie, 64
Stock
  brown bone, 131
  chicken, 132
  jellied, 132
  mixed, 133
  vegetable, 132
  white bone, 132
Strawberry
  charlotte, 99
  purée, 98
  walnut cream cake, 109

Tangerine sorbet, 99
Terrine
  chicken, 89
  duck, 89
  hare, 94
Tomato(es)
  sauce, 50
  to skin and seed, 140
Tongue, pressed, 93

Vanilla cream, 103
Veal
  blanquette, 70
  ossi buchi, 69
Vegetable bortsch, 38
Zabaione, 101